Lecture Notes in Mathematics

For information about Vols. 1–489, please contact your bookseller or Springer-Verlag.

Vol. 490: The Geometry of Metric and Linear Spaces. Proceedings 1974. Edited by L. M. Kelly. X, 244 pages. 1975.

Vol. 491: K. A. Broughan, Invariants for Real-Generated Uniform Topological and Algebraic Categories. X, 197 pages. 1975.

Vol. 492: Infinitary Logic: In Memoriam Carol Karp. Edited by D. W. Kueker. VI, 206 pages. 1975.

Vol. 493: F. W. Kamber and P. Tondeur, Foliated Bundles and Characteristic Classes. XIII, 208 pages. 1975.

Vol. 494: A Cornea and G. Licea. Order and Potential Resolvent Families of Kernels. IV, 154 pages. 1975.

Vol. 495: A. Kerber, Representations of Permutation Groups II. V, 175 pages. 1975.

Vol. 496: L. H. Hodgkin and V. P. Snaith, Topics in K-Theory. Two Independent Contributions. III, 294 pages. 1975.

Vol. 497: Analyse Harmonique sur les Groupes de Lie. Proceedings 1973–75. Edité par P. Eymard et al. VI, 710 pages. 1975.

Vol. 498: Model Theory and Algebra. A Memorial Tribute to Abraham Robinson. Edited by D. H. Saracino and V. B. Weispfenning. X, 463 pages. 1975.

Vol. 499: Logic Conference, Kiel 1974. Proceedings. Edited by G. H. Müller, A. Oberschelp, and K. Potthoff. V, 651 pages 1975.

Vol. 500: Proof Theory Symposion, Kiel 1974. Proceedings. Edited by J. Diller and G. H. Müller. VIII, 383 pages. 1975.

Vol. 501: Spline Functions, Karlsruhe 1975. Proceedings. Edited by K. Böhmer, G. Meinardus, and W. Schempp. VI, 421 pages. 1976.

Vol. 502: János Galambos, Representations of Real Numbers by Infinite Series. VI, 146 pages. 1976.

Vol. 503: Applications of Methods of Functional Analysis to Problems in Mechanics. Proceedings 1975. Edited by P. Germain and B. Nayroles. XIX, 531 pages. 1976.

Vol. 504: S. Lang and H. F. Trotter, Frobenius Distributions in GL_2-Extensions. III, 274 pages. 1976.

Vol. 505: Advances in Complex Function Theory. Proceedings 1973/74. Edited by W. E. Kirwan and L. Zalcman. VIII, 203 pages. 1976.

Vol. 506: Numerical Analysis, Dundee 1975. Proceedings. Edited by G. A. Watson. X, 201 pages. 1976.

Vol. 507: M. C. Reed, Abstract Non-Linear Wave Equations. VI, 128 pages. 1976.

Vol. 508: E. Seneta, Regularly Varying Functions. V, 112 pages. 1976.

Vol. 509: D. E. Blair, Contact Manifolds in Riemannian Geometry. VI, 146 pages. 1976.

Vol. 510: V. Poènaru, Singularités C^∞ en Présence de Symétrie. V, 174 pages. 1976.

Vol. 511: Séminaire de Probabilités X. Proceedings 1974/75. Edité par P. A. Meyer. VI, 593 pages. 1976.

Vol. 512: Spaces of Analytic Functions, Kristiansand, Norway 1975. Proceedings. Edited by O. B. Bekken, B. K. Øksendal, and A. Stray. VIII, 204 pages. 1976.

Vol. 513: R. B. Warfield, Jr. Nilpotent Groups. VIII, 115 pages. 1976.

Vol. 514: Séminaire Bourbaki vol. 1974/75. Exposés 453 – 470. IV, 276 pages. 1976.

Vol. 515: Bäcklund Transformations. Nashville, Tennessee 1974. Proceedings. Edited by R. M. Miura. VIII, 295 pages. 1976.

Vol. 516: M. L. Silverstein, Boundary Theory for Symmetric Markov Processes. XVI, 314 pages. 1976.

Vol. 517: S. Glasner, Proximal Flows. VIII, 153 pages. 1976.

Vol. 518: Séminaire de Théorie du Potentiel, Proceedings Paris 1972–1974. Edité par F. Hirsch et G. Mokobodzki. VI, 275 pages. 1976.

Vol. 519: J. Schmets, Espaces de Fonctions Continues. XII, 150 pages. 1976.

Vol. 520: R. H. Farrell, Techniques of Multivariate Calculation. X, 337 pages. 1976.

Vol. 521: G. Cherlin, Model Theoretic Algebra – Selected Topics. IV, 234 pages. 1976.

Vol. 522: C. O. Bloom and N. D. Kazarinoff, Short Wave Radiation Problems in Inhomogeneous Media: Asymptotic Solutions. V. 104 pages. 1976.

Vol. 523: S. A. Albeverio and R. J. Høegh-Krohn, Mathematical Theory of Feynman Path Integrals. IV, 139 pages. 1976.

Vol. 524: Séminaire Pierre Lelong (Analyse) Année 1974/75. Edité par P. Lelong. V, 222 pages. 1976.

Vol. 525: Structural Stability, the Theory of Catastrophes, and Applications in the Sciences. Proceedings 1975. Edited by P. Hilton. VI, 408 pages. 1976.

Vol. 526: Probability in Banach Spaces. Proceedings 1975. Edited by A. Beck. VI, 290 pages. 1976.

Vol. 527: M. Denker, Ch. Grillenberger, and K. Sigmund, Ergodic Theory on Compact Spaces. IV, 360 pages. 1976.

Vol. 528: J. E. Humphreys, Ordinary and Modular Representations of Chevalley Groups. III, 127 pages. 1976.

Vol. 529: J. Grandell, Doubly Stochastic Poisson Processes. X, 234 pages. 1976.

Vol. 530: S. S. Gelbart, Weil's Representation and the Spectrum of the Metaplectic Group. VII, 140 pages. 1976.

Vol. 531: Y.-C. Wong, The Topology of Uniform Convergence on Order-Bounded Sets. VI, 163 pages. 1976.

Vol. 532: Théorie Ergodique. Proceedings 1973/1974. Edité par J.-P. Conze and M. S. Keane. VIII, 227 pages. 1976.

Vol. 533: F. R. Cohen, T. J. Lada, and J. P. May, The Homology of Iterated Loop Spaces. IX, 490 pages. 1976.

Vol. 534: C. Preston, Random Fields. V, 200 pages. 1976.

Vol. 535: Singularités d'Applications Differentiables. Plans-sur-Bex. 1975. Edité par O. Burlet et F. Ronga. V, 253 pages. 1976.

Vol. 536: W. M. Schmidt, Equations over Finite Fields. An Elementary Approach. IX, 267 pages. 1976.

Vol. 537: Set Theory and Hierarchy Theory. Bierutowice, Poland 1975. A Memorial Tribute to Andrzej Mostowski. Edited by W. Marek, M. Srebrny and A. Zarach. XIII, 345 pages. 1976.

Vol. 538: G. Fischer, Complex Analytic Geometry. VII, 201 pages. 1976.

Vol. 539: A. Badrikian, J. F. C. Kingman et J. Kuelbs, Ecole d'Eté de Probabilités de Saint Flour V-1975. Edité par P.-L. Hennequin. IX, 314 pages. 1976.

Vol. 540: Categorical Topology, Proceedings 1975. Edited by E. Binz and H. Herrlich. XV, 719 pages. 1976.

Vol. 541: Measure Theory, Oberwolfach 1975. Proceedings. Edited by A. Bellow and D. Kölzow. XIV, 430 pages. 1976.

Vol. 542: D. A. Edwards and H. M. Hastings, Čech and Steenrod Homotopy Theories with Applications to Geometric Topology. VII, 296 pages. 1976.

Vol. 543: Nonlinear Operators and the Calculus of Variations, Bruxelles 1975. Edited by J. P. Gossez, E. J. Lami Dozo, J. Mawhin, and L. Waelbroeck, VII, 237 pages. 1976.

Vol. 544: Robert P. Langlands, On the Functional Equations Satisfied by Eisenstein Series. VII, 337 pages. 1976.

Vol. 545: Noncommutative Ring Theory. Kent State 1975. Edited by J. H. Cozzens and F. L. Sandomierski. V, 212 pages. 1976.

Vol. 546: K. Mahler, Lectures on Transcendental Numbers. Edited and Completed by B. Diviš and W. J. Le Veque. XXI, 254 pages. 1976.

Vol. 547: A. Mukherjea and N. A. Tserpes, Measures on Topological Semigroups: Convolution Products and Random Walks. V, 197 pages. 1976.

Vol. 548: D. A. Hejhal, The Selberg Trace Formula for PSL (2, ℝ). Volume I. VI, 516 pages. 1976.

Vol. 549: Brauer Groups, Evanston 1975. Proceedings. Edited by D. Zelinsky. V, 187 pages. 1976.

Vol. 550: Proceedings of the Third Japan – USSR Symposium on Probability Theory. Edited by G. Maruyama and J. V. Prokhorov. VI, 722 pages. 1976.

continuation on page 153

Lecture Notes in Mathematics

Edited by A. Dold and B. Eckmann

727

Yoshimi Saitō

Spectral Representations for Schrödinger Operators With Long-Range Potentials

---

Springer-Verlag
Berlin Heidelberg New York 1979

Author

Yoshimi Saitō
Department of Mathematics
Osaka City University
Sugimoto-cho, Sumiyoshi-ku
Osaka/Japan

AMS Subject Classifications (1970): 35 J 10, 35 P 25, 47 A 40

ISBN 3-540-09514-4 Springer-Verlag Berlin Heidelberg New York
ISBN 0-387-09514-4 Springer-Verlag New York Heidelberg Berlin

Library of Congress Cataloging in Publication Data
Saitō, Yoshimi. Spectral representations for Schrödinger operators with long-range potentials. (Lecture notes in mathematics ; 727) Bibliography: p. Includes index. 1. Differential equations, Elliptic. 2. Schrödinger operator. 3. Scattering (Mathematics) 4. Spectral theory (Mathematics) I. Title. II. Series: Lecture notes in mathematics (Berlin) ; 727. QA3.L28 no. 727 [QA377] 510'.8s [515'.353] 79-15958

Printing and binding: Beltz Offsetdruck, Hemsbach/Bergstr.
2141/3140-543210

PREFACE

The present lecture notes are based on the lectures given at the University of Utah during the fall quarter of 1978. The main purpose of the lectures was to present a complete and self-contained exposition of the spectral representation theory for Schrödinger operators with long-range potentials.

I would like to thank Professor Calvin H. Wilcox of the University of Utah for not only the opportunity to present these lectures but also his unceasing encouragement and helpful suggestions. I would like to thank Professor Willi Jäger of the University of Heidelberg, too, for useful discussions during the lectures.

Yoshimi Saitō

TABLE OF CONTENTS

INTRODUCTION

In this series of lectures we shall be concerned with the spectral representations for Schrödinger operators

$$T = -\Delta + Q(y) \tag{0.1}$$

on the N-dimensional Euclidean space $\mathbf{R}^N$. Here

$$\begin{cases} y = (y_1, y_2, \ldots, y_N) \in \mathbf{R}^N , \\ \Delta = \sum_{j=1}^{N} \frac{\partial^2}{\partial y_j^2} \quad \text{(Laplacian)} , \end{cases} \tag{0.2}$$

and $Q(y)$, which is usually called a "potential", is a real-valued function on $\mathbf{R}^N$.

Let us explain the notion of spectral representation by recalling the usual Fourier transforms. The Fourier transforms $F_{0\pm}$ are defined by

$$(F_{0\pm}f)(\xi) = (2\pi)^{-\frac{N}{2}} \operatorname*{l.i.m.}_{R\to\infty} \int_{|y|<R} e^{\mp iy\xi} f(y)dy \quad \text{in } L_2(\mathbf{R}^N_\xi) , \tag{0.3}$$

where $f \in L_2(\mathbf{R}^N_y)$, $\xi = (\xi_1, \xi_2, \ldots, \xi_N) \in \mathbf{R}^N_\xi$, l.i.m. means the limit in the mean and $y\xi$ is the inner product in $\mathbf{R}^N$, i.e.,

$$y\xi = \sum_{j=1}^{N} y_j \xi_j . \tag{0.4}$$

Then it is well-known that $F_{0\pm}$ are unitary operators from $L_2(\mathbf{R}^N_y)$ onto $L_2(\mathbf{R}^N_\xi)$ and that the inverse operators $F^*_{0\pm}$ are given by

$$(F^*_{0\pm}F)(y) = (2\pi)^{-\frac{N}{2}} \underset{R\to\infty}{\text{l.i.m.}} \int_{|\xi|<R} e^{\pm iy\xi}F(\xi)d\xi \tag{0.5}$$

in $L_2(\mathbf{R}^N_y)$

for $F \in L_2(\mathbf{R}^N_\xi)$. Here it should be remarked that $e^{\pm iy\xi}$ are solutions of the equation

$$(-\Delta-|\xi|^2)u = 0 \qquad (|\xi|^2 = \sum_{j=1}^{N} \xi_j^2) \quad , \tag{0.6}$$

that is, $e^{\pm iy\xi}$ are the eigenfunctions (in a generalized sense) with the eigenvalue $|\xi|^2$ associated with $-\Delta$. Further, when $f(y)$ is sufficiently smooth and rapidly decreasing, we have

$$\begin{cases} F_{0\pm}(-\Delta f)(\xi) = |\xi|^2(F_{0\pm}f)(\xi) \\ \text{or} \\ (-\Delta f)(y) = F^*_{0\pm}(|\xi|^2F_{0\pm}f)(y) \quad , \end{cases} \tag{0.7}$$

which means that the partial differential operator $-\Delta$ is transformed into the multiplication operator $|\xi|^2\times$ by the Fourier transforms $F_{0\pm}$.

Now we can find deeper and clearer relations between the operator $-\Delta$ and the Fourier transforms if we consider the self-adjoint realization of the Laplacian in $L_2(\mathbf{R}^N)$. Let us define a symmetric operator h_0 in $L_2(\mathbf{R}^N)$ by

$$\begin{cases} \mathcal{D}(h_0) = C_0^\infty(\mathbf{R}^N) \\ h_0 f = -\Delta f \quad , \end{cases} \tag{0.8}$$

where $\mathcal{D}(h_0)$ is the domain of h_0 . Then, as is well-known, h_0 is essentially self-adjoint and its closure $H_0 = \tilde{h}_0$ is a unique self-adjoint extension of h_0 . H_0 is known to satisfy

$$
(0.9)\quad \begin{cases} \mathcal{D}(H_0) = \{f \in L_2(\mathbf{R}^N_y) / |\xi|^2(\mathcal{F}_{0\pm}f)(\xi) \in L_2(\mathbf{R}^N_\xi)\} \quad (= H_2(\mathbf{R}^N)) , \\ H_0 f = -\Delta f \qquad (f \in \mathcal{D}(H_0)) , \\ \mathcal{F}_{0\pm}(H_0 f)(\xi) = |\xi|^2(\mathcal{F}_{0\pm}f)(\xi) \quad (f \in \mathcal{D}(H_0)) , \\ E_0(B) = \mathcal{F}^*_{0\pm}\ \chi_{\sqrt{B}}\ \ \mathcal{F}_{0\pm} \quad , \end{cases}
$$

where $H_2(\mathbf{R}^N)$ is the Sobolev space of the order 2, B is an interval in $(0,\infty)$, $\chi_{\sqrt{B}}$ is the characteristic function of $\sqrt{B} = \{\xi \in \mathbf{R}^N_\xi / |\xi|^2 \in B\}$, and $E_0(\cdot)$ denotes the spectral measure associated with H_0 . The differential operator $-\Delta$ is acting on functions of $H_2(\mathbf{R}^N)$ in the distribution sense. Thus it can be seen that the Fourier transforms $\mathcal{F}_{0\pm}$ give unitary equivalence between H_0 and the multiplication operator $|\xi|^2 \times$.

Let us next consider the Schrödinger operator T . Throughout this lecture $\mathcal{Q}(y)$ is assumed to be a real-valued, continuous function on $\mathbf{R}^N$ and satisfy

$$
(0.10)\qquad \mathcal{Q}(y) \to 0 \qquad (|y| \to \infty) .
$$

Then a symmetric operator h defined by

$$
(0.11)\quad \begin{cases} \mathcal{D}(h) = C_0^\infty(\mathbf{R}^N) \quad , \\ hf = Tf \end{cases}
$$

can be easily seen to be essentially self-adjoint with a unique extension H in $L_2(\mathbf{R}^N)$. We have

$$
(0.12)\quad \begin{cases} \mathcal{D}(H) = \mathcal{D}(H_0) \\ Hf = Tf \qquad (f \in \mathcal{D}(H)) . \end{cases}
$$

Here the differential operator T in (0.12) should be taken in the distribution sense as in (0.9). These facts will be shown in §12 (Theorem 12.1). Our final aim is to construct the "generalized" Fourier transforms $F_{\pm}$ which are unitary operators from $L_2(\mathbf{R}_y^N)$ onto $L_2(\mathbf{R}_\xi^N)$ and satisfy

$$(0.13)\quad \begin{cases} \mathcal{D}(H) = \{f \in L_2(\mathbf{R}_y^N) / |\xi|^2 (F_{\pm} f)(\xi) \in L^2(\mathbf{R}_\xi^N)\} \ , \\ F_{\pm}(Hf)(\xi) = |\xi|^2 (F_{\pm} f)(\xi) \qquad (f \in \mathcal{D}(H)) \ , \\ E(B) = F_{\pm}^* \, \chi_{\sqrt{B}} \, F_{\pm} \ , \end{cases}$$

where B and $\chi_{\sqrt{B}}$ are as in (0.9) and $E(\cdot)$ is the spectral measure associated with H. It will be also shown that the generalized Fourier transforms $F_{\pm}$ are constructed by the use of solutions of the equation $Tu = |\xi|^2 u$.

Now let us state the exact conditions on the potential $\mathcal{Q}(y)$ in this lecture. Let us assume that $\mathcal{Q}(y)$ satisfies

$$(0.14)\qquad \mathcal{Q}(y) = o(|y|^{-\varepsilon}) \qquad (|y| \to \infty)$$

with a constant $\varepsilon > 0$. When $\varepsilon > 1$ the potential $\mathcal{Q}(y)$ is called a short-range potential and when $0 < \varepsilon \leqq 1$ $\mathcal{Q}(y)$ is called a long-range potential. The more rapidly diminishes $\mathcal{Q}(y)$ at infinity, the nearer becomes the Schrödinger operator T to $-\Delta$ and T can be treated the easier. In this lecture a spectral representation theorem will be shown for the Schrödinger operator T with a long-range potential $\mathcal{Q}(y)$ with additional conditions

$$(0.15)\qquad (D^j \mathcal{Q})(y) = o(|y|^{-j-\varepsilon}) \qquad (|y| \to \infty \ , \ j = 0,1,2, \ldots ,m_0) \ ,$$

where D^j is an arbitrary derivative of j-th order and m_0 is a constant determined by $\mathcal{Q}(y)$. Without these additional conditions the spectral structure of the self-adjoint realization H of T may be too different from the one of the self-adjoint realization H_0 of $-\Delta$. The Schrödinger operator with a potential of the form

$$\mathcal{Q}(y) = \frac{1}{|y|} \sin |y| \quad , \tag{0.16}$$

for example, is beyond the scope of this lecture.

The core of the method in this lecture is to transform the Schrödinger operator T into an ordinary differential operator with operator-valued coefficients. Let $I = (0,\infty)$ and $X = L_2(S^{N-1})$, S^{N-1} being the $(N-1)$-sphere, with its inner product $(\ ,\ )_X$ and norm $|\ |_X$. Let us introduce a Hilbert space $L_2(I,X)$ which is all X-valued functions $u(r)$ on $I = (0,\infty)$ satisfying

$$\|u\|_0^2 = \int_0^\infty |u(r)|_X^2 \, dr < \infty \quad . \tag{0.17}$$

Its inner product $(\ ,\ )_0$ is defined by

$$(u,v)_0 = \int_0^\infty (u(r)\ ,\ v(r))_X \, dr \quad . \tag{0.18}$$

Then a multiplication operator $U = r^{\frac{N-1}{2}} \times$ defined by

$$(0.19) \qquad \begin{aligned} &L_2(\mathbb{R}^N) \ni f(y) \quad \to r^{\frac{N-1}{2}} f(r\omega) \in L_2(I,X) \\ &\qquad\qquad (r=|y|\ ,\ \omega = \frac{y}{|y|} \in S^{N-1}) \end{aligned}$$

can be easily seen to be a unitary operator from $L_2(\mathbb{R}^N)$ onto $L_2(I,X)$.

Let L be an ordinary differential operator defined by

$$(0.20)\qquad L = -\frac{d^2}{dr^2} + B(r) + C(r) \qquad (r \in I) \quad ,$$

where, for each $r > 0$,

$$(0.21)\qquad \begin{cases} (C(r)x)(\omega) = Q(r\omega)x(\omega) & (x \in X) \\ (B(r)x)(\omega) = \frac{1}{r^2}\left(-(\Lambda_N x)(\omega) + \frac{(N-1)(N-3)}{4}\, x(\omega)\right) & \\ & (x \in \mathcal{D}(\Lambda_N)) \end{cases}$$

and Λ_N is a non-positive self-adjoint operator called the Laplace-Beltrami operator on S^{N-1} . If polar coordinates $(r,\theta_1,\theta_2, \ldots ,\theta_{N-1})$ are introduced by

$$(0.22)\qquad \begin{cases} y_j = r\sin\theta_1 \sin\theta_2 \ldots \sin\theta_{j-1} \cos\theta_j \\ \qquad\qquad (j = 1,2,\ldots,N-1) \quad , \\ y_N = r\sin\theta_1 \sin\theta_2 \ldots \sin\theta_{N-2} \sin\theta_{N-1} \\ \qquad (r \geqslant 0 \;,\; 0 \leqslant \theta_1,\theta_2,\ldots\theta_{N-2} \leqslant 2\pi \;,\; 0 \leqslant \theta_{N-1} < 2\pi) \;, \end{cases}$$

then $\Lambda_N x$ can be written as

$$(0.23)\qquad \begin{aligned} &(\Lambda_N x)(\theta_1,\theta_2,\ldots,\theta_{N-1}) \\ &= \sum_{j=1}^{N-1} (\sin\theta_1\ldots\sin\theta_{j-1})^{-2}(\sin\theta_j)^{-N+j+1} \frac{\partial}{\partial\theta_j}\left\{(\sin\theta_j)^{N-j-1} \frac{\partial x}{\partial\theta_j}\right\} \end{aligned}$$

for $x \in C^2(S^{N-1})$ (see, e.g., Erdélyi and al. [1], p. 235). The operator (0.23) with the domain $C^2(S^{N-1})$ is essentially self-adjoint and its closure is Λ_N. The operators T and L are combined by the use of the unitary operator U in the following way

$$T\Phi = U^{-1}LU\Phi \tag{0.24}$$

for an arbitrary smooth function Φ on $\mathbb{R}^N$. Thus, for the time being, we shall be concerned with L instead of T, and after that the results obtained for L will be translated into the results for T.

Our operator T corresponds to the two-particle problem on the quantum mechanics and there are many works on the spectral representation theorem for T with various degrees of the smallness assumption on the potential $\mathcal{Q}(y)$. Let $\mathcal{Q}(y)$ satisfy

$$\mathcal{Q}(y) = O(|y|^{-\varepsilon}) \qquad (|y| \to \infty, \quad \varepsilon > 0). \tag{0.25}$$

In 1960, Ikebe [1] developed an eigenfunction expansion theory for T in the case of spatial dimension $N = 3$ and $\varepsilon > 2$. He defined a generalized eigenfunction $\Phi(y,\xi)$ $(y,\xi \in \mathbb{R}^3)$ as the solution of the Lippman-Schwinger equation

$$\Phi(y,\xi) = e^{iy\xi} - \frac{1}{4\pi}\int_{\mathbb{R}^3} \frac{e^{i|\xi||y-z|}}{|y-z|}\mathcal{Q}(z)\Phi(z,\xi)dz, \tag{0.26}$$

and by the use of $\Phi(y,\xi)$, he constructed the generalized Fourier transforms associated with T. The results of Ikebe [1] were extended by Thoe [1] and Kuroda [1] to the case where N is arbitrary and $\varepsilon > \frac{1}{2}(N + 1)$. On the other hand, Jäger [1]-[4] investigated the properties of an ordinary differential operator with operator-valued coefficients and obtained a spectral representation theorem for it. His results can be applied to the case of $N \geqslant 3$ and $\varepsilon > \frac{3}{2}$. Saitō [1],[2] extended the results of Jäger to include the case of $N \geqslant 3$ and $\varepsilon > 1$, i.e., the general short-range case. At almost the same time S. Agmon obtained an eigenfunction expansion theorem for general elliptic operators in $\mathbb{R}^N$ with short-range coefficients. His results can be seen in Agmon [1]. Thus, the short-range case has been settled. As for the long-range case, after the work of Dollard [1] which deals with Coulomb potential $\mathcal{Q}(y) = \frac{c}{|y|}$, Saitō [3]-[5] treated the case of $\varepsilon > \frac{1}{2}$ along the line of Saitō [1]-[2]. Ikebe [2],[3] treated the Schrödinger operator with $\varepsilon > \frac{1}{2}$ directly using essentially the same ideas as Saitō [3]-[5]. After these works, Saitō [6],[7] showed a spectral representation theorem for the Schrödinger operator with a general long-range potential $\mathcal{Q}(y)$ with $\varepsilon > 0$. S. Agmon also obtains an eigenfunction expansion theorem for general elliptic operators with long-range coefficients.

Let us outline the contents of this lecture. Throughout this lecture, the spatial dimension N is assume to be $N \geqslant 3$. Then we have $B(r) \geqslant 0$ for each $r > 0$. As for the case of $N = 2$, see Saitō [6], §5.

In Chapter I we shall show the limiting absorption principle for L which enables us to solve the equation

$$(0.27) \qquad \begin{cases} (L - k^2)v = f, \\ \\ (\frac{d}{dr} - ik)v \to 0 \quad (r \to \infty), \end{cases}$$

for not only k^2 non-real, but also k^2 real.

In Chapter II, the asymptotic behavior of $v(r)$, the solution of the equation (0.27) with real k, will be investigated. When $\mathcal{Q}(y)$ is short-range, there exists an element $x_v \in X$ such that the asymptotic relation

$$(0.28) \qquad v(r) \sim e^{irk}x_v \quad (v \to \infty)$$

holds. But, in the case of the long-range potential, (0.28) should be modified in the following

$$v(r) \sim e^{irk - i\lambda(r\cdot,k)}\tilde{x}_v \tag{0.29}$$

with $\tilde{x}_v \in X$. Here the function $\lambda(y,k)$, which will be constructed in Chapter II, is called a <u>stationary modifier</u>.

This asymptotic formula (0.29) will play an important role in spectral representation theory for L which will be developed in Chapter III.

The contents of this lecture are essentially given in the papers of Saitō [3]-[7], though they will be developed into a more unified and self-contained form in this lecture. But we have to assume in advance the elements of functional analysis and theory of partial differential equations: to be more precise, the elemental properties of the Hilbert space and the Banach space, spectral decomposition theorem of self-adjoint operators, elemental knowledge of distribution, etc.

CHAPTER I
LIMITING ABSORPTION PRINCIPLE

§1. Preliminaries.

Let us begin with introducing some notations which will be employed without further reference.

$\mathbb{R}$: real numbers

$\mathbb{C}$: complex numbers

$\mathbb{C}^+ = \{k = k_1 + ik_2 \in \mathbb{C}/k_1 \neq 0,\ k_2 \geqslant 0\}$.

Rek: the real part of k.

Imk: the imaginary part of k.

$I = (0, \infty) = \{r \in \mathbb{R}/\ 0 < r < \infty\}$.

$\bar{I} = [0, \infty) = \{r \in \mathbb{R}/\ 0 \leqq r < \infty\}$.

$X = L_2(S^{N-1})$ with its norm $|\ |_X$ and inner product $(\ ,\ )_X$

$L_{2,s}(J,X)$, $s \in \mathbb{R}$, is the Hilbert space of all X-valued functions $f(r)$ on an interval J such that $(1+r)^s|f(r)|_X$ is square integrable on J. The inner product $(\ ,\ )_{s,J}$ and norm $\|\ \|_{s,J}$ are defined by

$$(f,g)_{s,J} = \int_J (1+r)^{2s}(f(r),\ g(r))_X dr$$

and

$$\|f\|_{s,J} = [(f,f)_{s,J}]^{\frac{1}{2}} ,$$

respectively. When $s = 0$ or $J = I$ the subscript 0 or I may be omitted as in $L_2(J,X)$, $\|\ \|_s$ etc.

$C_0^\infty(J,X) = UC_0^\infty(\Omega_J)$, where J is an open interval in I,

$\Omega_J = \{y \in \mathbb{R}^N/|y| \in J\}$ and $U = r^{\frac{N-1}{2}}x$ is given by (0,18).

$H_{0,s}^{1,B}(J,X)$, $s \in \mathbb{R}$, is the Hilbert space obtained by the completion of a pre-Hilbert space $C_0^\infty(J\ X)$ with its norm

$$\|\phi\|_{B,s,J} = \left[\int_J (1+r)^{2s}\left\{|\phi'(r)|_X^2 + |B^{\frac{1}{2}}(r)\phi(r)|_X^2 + |\phi(r)|_X^2\right\}dr\right]^{\frac{1}{2}}$$

and inner product

$$(\phi,\psi)_{B,s,J} = \int_J (1+r)^{2s}\Big\{(\phi'(r),\psi'(r))_X + (B^{\frac{1}{2}}(r)\phi(r),\ B^{\frac{1}{2}}(r)\psi(r))_X + (\phi(r),\psi(r))_X\Big\}dr\ ,$$

where $B^{\frac{1}{2}}(r) = (B(r))^{\frac{1}{2}}$ with $B(r)$ given by (0.21) and $\phi'(r) = \frac{\partial\phi}{\partial r}$. When $J = I$ or $s = 0$ we shall omit the subscript I or s as in $\|\ \|_B$, $H_0^{1,B}(I,X)$ etc.

$F_\beta(J,X)$, $\beta \geq 0$, is the set of all anti-linear continuous functional ℓ on $H_0^{1,B}(J,X)$, i.e.,

$$\ell\colon H_0^{1,B}(J,X) \ni v \to \langle\ell,v\rangle \in \mathbb{C}\ ,$$

such that

$$|||\,\ell\,|||_{\beta,J} = \sup\{|\langle\ell,(1+r)^\beta\phi\rangle| / \phi \in C_0^\infty(J,X)\ ,\ \|\phi\|_{B,J} = 1\} < \infty.$$

$F_\beta(J,X)$ is a Banach space with its norm $|||\ |||_{\beta,J}$. When $\beta = 0$ or $J = I$ the subscript 0 or I will be omitted as in $F(J,X)$, $|||\ \ |||_J$, $|||\ \ |||_\beta$ etc.

$C_{ac}(J,X)$, J being an (open or closed) interval, is the set of all X-valued functions $f(r)$ on J such that $f(r)$ is strongly absolutely continuous on every compact interval in J and there exists the weak derivative $f'(r)$ for almost all $r \in J$ with $f'(r) \in L_2(J',X)$ for any compact interval $J' \subset J$.

$C^j(J,X)$ is the set of all X-valued functions on J having j strong continuous derivatives. Here j is a non-negative integer and J is an (open or closed) interval.

$L_2(I,X)_{loc}$ $(H_0^{1,B}(I,X)_{loc})$ is the set of all X-valued functions $f(r)$ such that $\rho f \in L_2(I,X)$ $(\rho f \in H_0^{1,B}(I,X))$ for any real-valued C^1 function $\rho(r)$ on $\bar{I}$ having a compact support in $\bar{I}$.

$\mathbb{B}(Y,Z)$ is the Banach space of all bounded linear operators from Y into Z with its operator norm $\| \ \|_{Y,Z}$, where Y and Z are Banach spaces. We set $\mathbb{B}(Y,Y) = \mathbb{B}(Y)$ and the norm of $\mathbb{B}(X)$ is denoted simply by $\| \ \|$.

$\mathcal{D}(W)$ means the domain of W.

$A = A_N = -\Lambda_N + \frac{1}{4}(N-1)(N-3)$, where Λ_N is the Laplace-Beltrani operator (as a self-adjoint operator in X).

$D = \mathcal{D}(A)$.

$D^{\frac{1}{2}} = \mathcal{D}(A^{\frac{1}{2}})$.

$C(\alpha,\beta, \ldots)$ denotes a positive constant depending only on $\alpha, \beta, \ldots$ But very often symbols indicating obvious dependence will be omitted.

$H_j(\mathbb{R}^N)$ is the Sobolev space of the order j, i.e., the set of all L_2 functions with L_2 distribution derivatives up to the j-th order, inclusive.

$C_0^\infty(\mathbb{R}^N)$, $L_2(\Omega)$, $L_2(\Omega, (1+|y|)^{2s}dy)$ etc., will be employed as usual, where Ω is a measuable set in $\mathbb{R}^N$.

Let L be the differential operator defined by (0.20) and (0.21). The remainder of this section will be devoted to showing some basic properties of a (weak) solution of the equation $(L - k^2)v = f$. To this end, we shall first list some properties of $H_0^{1,B}(I, X)$.

PROPOSITION 1.1. (i) Let J be an open interval in I and let $\Omega_J = \{y \in \mathbb{R}^N / |y| \in J\}$. Then

$$(1.1) \qquad H_0^{1,B}(J,X) = UH_1(\Omega_J)$$

and

$$(1.2) \qquad (U\varphi,\ U\psi)_{B,J} = (\varphi,\psi)_{1,\Omega_J} \qquad (\varphi,\ \psi \in H_1(\Omega_J)),$$

where U is given by (0.18) and $(\ ,\ )_{1,\Omega_J}$ is the inner product of $H_1(\Omega_J)$, i.e.,

$$(1.3) \qquad (\varphi,\psi)_{1,\Omega_J} = \int_{\Omega_J} \left\{(\nabla\varphi)(y) \cdot \overline{(\nabla\psi)(y)} + \varphi(y)\overline{\psi(y)}\right\} dy.$$

(ii) Let $v \in H_0^{1,B}(I,X)$ (or $v \in H_0^{1,B}(I,X)_{loc}$). Then $v \in C_{ac}(\bar{I},X) \cap L_2(I,X)$ (or $v \in C_{ac}(\bar{I},X)$ $L_2(I,X)_{loc}$) with the weak derivative $v' \in L_2(I,X)$ (or $v' \in L_2(I,X)_{loc}$). Moreover $v(r) \in D^{\frac{1}{2}}$ for almost all $r \in I$ with $B^{\frac{1}{2}}v \in L_2(I,X)$ (or $B^{\frac{1}{2}}v \in L_2(I,X)_{loc}$). The inner product $(\ ,\ )_B$ and norm $\|\ \|_B$ of $H_0^{1,B}(I,X)$ have the following form:

$$(1.4) \qquad \begin{cases} (u,v)_B = \displaystyle\int_I \Big\{(u'(r),v'(r))_X + (B^{\frac{1}{2}}(r)u(r),B^{\frac{1}{2}}(r)v(r))_X \\ \qquad\qquad\qquad + (u(r),v(r))_X\Big\}dr, \\ \|u\|_B = \left[(u,u)_B\right]^{\frac{1}{2}} \end{cases}$$

(iii) Let $v \in H_0^{1,B}(I,X)_{loc}$. Then we have the relations

$$(1.5) \qquad v(0) = 0,$$

$$(1.6) \qquad |v(r) - v(s)|_X \leqq |r-s|^{\frac{1}{2}} \|v\|_{B,(0,R)} \qquad (r,s \in [0,R])$$

with

$$(1.7) \quad \|v\|_{B,(0,R)} = \left[\int_0^R \left\{|v'(r)|_X^2 + |B^{\frac{1}{2}}(r)v(r)|_X^2 + |v(r)|^2\right\} dr\right]^{\frac{1}{2}},$$

$$(1.8) \qquad |v(r)|_X \leqq \sqrt{2}\, \|v\|_{B,(0,R+1)} \qquad (r \in [0,R]) .$$

PROOF. (i) follows from two facts that $H_1(\Omega_J)$ is the Hilbert space obtained by completion of $C_0^\infty(\Omega_J)$ by the norm (1.3) and that the relation $\|U\varphi\|_{B,J} = \|\varphi\|_{1,\Omega_J}$ holds for $\varphi \in C_0^\infty(\Omega_J)$. The second fact can be easily obtained from the relation $(-\Delta)\varphi = U^{-1}L_0U\varphi$, where we set $L_0 = \frac{d^2}{dr^2} + B(r)$.

Let us show (ii) only for $v \in H_0^{1,B}(I,X)$, because the statement about $v \in H_0^{1,B}(I,X)_{loc}$ can be shown in quite a similar way. By the definition of $H_0^{1,B}(I,X)$ $v \in H_0^{1,B}(I,X)$ if and only if there exists a sequence $\{\phi_n\} \subset C_0^\infty(I,X)$ such that the sequences $\{\phi_n\}$, $\{\phi_n'\}$, $\{B^{\frac{1}{2}}\phi_n\}$ are Cauchy sequences in $L_2(I,X)$ and ϕ_n converges to v in $L_2(I,X)$. We set $\lim_{n\to\infty} \phi_n' = v_1$ and $\lim_{n\to\infty} B^{\frac{1}{2}}\phi_n = v_2$. From the estimate

$$(1.9) \qquad |\phi_n(r) - \phi_m(r)|_X \leqq \int_0^r |\phi_n'(t) - \phi_m'(t)|_X dt$$

it follows that the sequence $\{\phi_n(r)\}$ is convergent in X uniformly for $r \in [0,R]$ with an arbitrary positive R. Therefore $v(r) = \text{s-}\lim_{n\to\infty} \phi_n(r)$ is an X-valued, continuous function on $\overline{I}$. Letting $n \to \infty$ in the relation

$$(1.10) \quad (\phi_n(r)-\phi_n(s),x)_X = \int_s^r (\phi_n'(t),x)_X dt \qquad (x \in X,\ 0 \leqq s < r < \infty)$$

we arrive at

$$(1.11) \qquad (v(r) - v(s),\ x)_X = \int_s^r (v_1(t),x)_X dt,$$

whence follows that $v \in C_{ac}(\bar{I},X)$ and the weak derivative $v' = v_1$ in $L_2(I,X)$. Since $B^{\frac{1}{2}}\phi_n$ converges to v_2 in $L_2(I,X)$, there exists a null set e of I such that $B^{\frac{1}{2}}(r)\phi_n(r)$ converges to $v_2(r)$ in X for $r \in I - e$. Therefore $v(r) \in D^{\frac{1}{2}}$ with $B^{\frac{1}{2}}(r)v(r) = v_2(r)$ for $r \in I - e$ because of the closedness of the operator $B^{\frac{1}{2}}(r)$. Thus we obtain

$$(1.12) \quad \|v\|_B = \lim_{n\to\infty} \|\phi_n\|_B = \left[\int_I \left\{|v'(r)|_X^2 + |B^{\frac{1}{2}}(r)v(r)|_X^2 + |v(r)|_X^2\right\} dr\right]^{\frac{1}{2}}.$$

In a similar way we can show the first relation of (1.4) which is related the inner product $(u,v)_B$.

Finally, let us show (iii). Take a real-valued C^∞ function $\rho(r)$ on $\bar{I}$ such that $\rho(r) = I$ $(r \leqq R + 1)$, $= 0$ $(r \geqq R + 2)$. Then, since $\rho v \in H_0^{1,B}(I,X)$, there exists $\{\phi_n\} \subset C_0^\infty(I,X)$ such that $\|\rho v - \phi_n\|_B \to 0$ as $n \to \infty$, whence follows that

$$(1.13) \qquad \|v - \phi_n\|_{B,(0,R+1)} \to 0 \qquad (n \to \infty).$$

Then, as is seen in the proof of (ii), $\phi_n(r)$ converges $v(r)$ in X for $r \in [0, R + 1]$. (1.5) follows from the fact that $\phi_n(0) = 0$ for all $n = 1,2,\ldots$ (1.6) and (1.8) are obtained by letting $n \to \infty$ in the relations

$$(1.14) \qquad |\phi_n(r)-\phi_n(s)|_X = \left|\int_s^r \phi_n'(t)dt\right|_X \leqq |r-s|^{\frac{1}{2}} \|\phi_n\|_{B,(0,R)}$$

$$(0 \leqq r,\ s \leqq R)$$

and

$$(1.15)\qquad |\phi_n(r)|_X^2 \leqq 2(|\phi_n(r)-\phi_n(t)|_X^2 + |\phi_n(t)|_X^2)$$

$$\leqq 2(|r-t| \int_r^t |\phi_n'(s)|_X^2\, ds + \int_r^{r+1} |\phi_n(s)|_X^2\, ds)$$

$$\leqq 2\|\phi_n\|_{B,(0,R+1)}^2 \qquad (0 \leqq r \leqq R)$$

respectively, where $t \in [r, r+1]$ has been taken to satisfy

$|\phi_n(t)|_X = \min_{r \leqq s \leqq r+1} |\phi_n(s)|_X$. Q.E.D.

The interior estimate for the operator L is shown in the following:

PROPOSITION 1.2. Let L be as above and let $\mathcal{Q}(y)$ be a continuous function on $\mathbb{R}^N$. Let $v \in H_0^{1,B}(I,X)_{loc}$ satisfies the equation

$$(1.16)\qquad (v, (L-\bar{k}^2)\phi)_0 = \langle \ell, \phi \rangle \quad (\phi \in C_0^\infty(I,X))$$

with $k \in \mathbb{C}$ and $\ell \in F(I,X)$. Let R be a positive number. Then there exists $C = C(k,R)$ such that

$$(1.17)\qquad \|v\|_{B,(0,R)} \leqq C\left\{\|v\|_{0,(0,R+1)} + |||\ell|||_{0,(0,R+1)}\right\}.$$

The constant $C(k,R)$ is bounded when the pair (k,R) moves in a bounded set of $\mathbb{C} \times I$.

PROOF. Since $v \in H_0^{1,B}(I,X)_{loc}$, we have by (ii) of Proposition 1.1

$$(1.18)\qquad \begin{cases} \dfrac{d}{dr}(v(r), \phi'(r))_X = (v'(r), \phi'(r))_X + (v(r), \phi''(r))_X \\ \\ (v(r), B(r)\phi(r))_X = (B^{\frac{1}{2}}(r)v(r), B^{\frac{1}{2}}(r)\phi(r))_X \end{cases}$$

for almost all $r \in I$. Therefore from (1.16) it follows by partial integration that

$$(1.19)\qquad \int_0^{R+1}\Big\{(v'(r),\phi'(r))_X + (B^{\frac{1}{2}}(r)v(r),\ B^{\frac{1}{2}}(r)\phi(r))_X + (v(r),\ (C(r)-\overline{k}^2)\phi(r))_X\Big\}dr = \langle \ell,\ \phi\rangle$$

for any $\phi \in C_0^\infty(I,X)$. Set $\phi = \psi^2\phi_n$ in (1.19), where $\phi_n \in C_0^\infty((0,R+2),X)$ such that $\|v - \phi_n\|_{B,(0,R+1)} \to 0$ as $n \to \infty$ and $\psi(r)$ is a real-valued C^∞ function on $[0, R+2]$ such that $0 \leqq \psi \leqq 1$, $\psi(r) = 1$ $(0 \leqq r \leqq R)$, $\psi(r) = 0$ $(R+1 \leqq r \leqq R+2)$. Then, letting $n \to \infty$, we have

$$(1.20)\qquad \int_0^{R+1}\psi^2\Big\{|v'|_X^2 + |B^{\frac{1}{2}}v|_X^2 + |v|_X^2\Big\}dr + 2\int_0^R \psi\psi'(v',v)_X dr$$

$$= \int_0^{R+1}\psi^2(v,(\overline{k}^2 + 1 - C)v)_X dr + \langle \ell,\ \psi^2 v\rangle .$$

(1.17) is obtained from (1.20) by the use of the Schwarz inequality.

Q.E.D.

<u>PROPOSITION 1.3</u> (regularity theorem). Let L be as in Proposition 1.2. Let $v \in L_2(I,X)_{loc}$ be a "weak" solution of the equation $(L-k^2)v = f$ with $k \in \mathbb{C}^2$ and $f \in L_2(I,X)_{loc}$, i.e., let v satisfy

$$(1.21)\qquad (v,\ (L-\overline{k}^2)\phi)_0 = (f,\phi)_0$$

for all $\phi \in C_0^\infty(I,X)$. Then v satisfies following (1) ~ (4):

(1) $v \in H_0^{1,B}(I,X)_{loc} \cap C^1(I,X)$. $v(r) \in D^{\frac{1}{2}}$ for each $r \geqq 0$. $v(r) \in D$ for almost all $r \in I$ with $Bv \in L_2((a,b),X)$ for any $0 < a < b < \infty$.

(2) $v' \in C_{ac}(I,X)$ with the weak derivative v'' and $v'(r) \in D^{\frac{1}{2}}$ for almost all $r \in I$. $B^{\frac{1}{2}}v' \in L_2((a,b),X)$ for any $0 < a < b < \infty$.

(3) $B^{\frac{1}{2}}v \in C_{ac}(I,X)$ with the weak derivative $(B^{\frac{1}{2}}v)'$.

$$(B^{\frac{1}{2}}(r)v(r))' = -\frac{1}{r}B^{\frac{1}{2}}(r)v(r) + B^{\frac{1}{2}}(r)v'(r) \tag{1.22}$$

holds for almost all $r \in I$.

(4) We have

$$-v''(r) + B(r)v(r) + C(r)v(r) - k^2 v(r) = f(r) \tag{1.23}$$

for almost all $r \in I$.

PROOF. Set $\tilde{v} = U^{-1}v$ and $\tilde{f} = U^{-1}f$. Then $\tilde{v}$, $\tilde{f}$ belong to $L_2(\mathbb{R}^N)_{loc}$ and from (1.21) the relation

$$(\tilde{v}, (T - k^2)\varphi)_{L_2} = (\tilde{f}, \varphi)_{L_2} \qquad (\varphi \in C_0^\infty(\mathbb{R}^N)) \tag{1.24}$$

follows, where $(\ ,\ )_{L_2}$ denotes the inner product of $L_2(\mathbb{R}^N)$. As is well-known, (1.24) implies that $\tilde{v} \in H_2(\mathbb{R}^N)_{loc}$ (see, e.g., Ikebe-Kato [1]). Therefore there exists a sequence $\{\varphi_n\} \subset C_0^\infty(\mathbb{R}^N)$ such that $\varphi_n \to \tilde{v}$ in $H_2(\mathbb{R}^N)_{loc}$ as $n \to \infty$. Set $\phi_n = U\varphi_n$. Then $\phi_n \to v$ in $H_0^{1,B}(I,X)_{loc}$ as $n \to \infty$ and $\{\phi_n''\}$, $\{B\phi_n\}$, $\{B^{\frac{1}{2}}\phi_n'\}$ are all Cauchy sequences in $L_2((a,b),X)$ $(0 < a < b < \infty)$. From these relations we can easily obtain (1) ~ (4).

Q.E.D.

Now that Proposition 1.3 has been established, let us introduce one more function space. Let J be an open interval in I. $D(J)$ denotes the set of all functions v on J which satisfy the following (a) and (b):

(a) $v \in C^1(J,X)$. $v(r) \in D^{\frac{1}{2}}$ for each $r \in J$. For almost all $r \in J$ $v(r) \in D$ and $v'(r) \in D^{\frac{1}{2}}$. Bv and $B^{\frac{1}{2}}v'$ belong to $L_2(J',X)$ for any compact interval J' in J.

(b) v', $B^{\frac{1}{2}}v \in C_{ac}(J,X)$ with

$$(B^{\frac{1}{2}}v)' = -\frac{1}{r}B^{\frac{1}{2}}v + B^{\frac{1}{2}}v'$$

for almost all $r \in J$.

It follows from Proposition 1.3 that the solution $v \in L_2(I,X)_{loc}$ of the equation (1.21) belongs to $H_0^{1,B}(I,X)_{loc} \cap D(I)$.

The unique continuation theorem for the operator L takes the following form.

PROPOSITION 1.4. Let L be as in Proposition 1.2 and let $v \in D(J)$ satisfy the equation $(L-k^2)v = 0$ on J with $k \in \mathbb{C}$, where J is an open interval in I. Suppose that $v(r) = 0$ in a neighborhood of some point $r_0 \in J$. Then $v(r) = 0$ on J.

PROOF. Set $\tilde{v} = U^{-1}v$ as in the proof of Proposition 1.3. Then $\tilde{v}$ is a solution of the equation $(T-k^2)\tilde{v} = 0$ and $\tilde{v}(y) = 0$ for $y \in \{y \in \mathbb{R}^N / ||y| - r_0| < \eta\}$ with some $\eta > 0$. Thus we can apply the unique continuation theorem for elliptic operators (see, e.g., Aronszajn [1]) to show that $\tilde{v} \equiv 0$ on $\{y \in \mathbb{R}^N / |y| \in J\}$, which completes the proof.

Q.E.D.

Finally we shall show a proposition which is a version of the Relich theorem.

PROPOSITION 1.5. Let L be as in Proposition 1.2 and let J be a bounded open interval in I. Let $\{v_n\}$ be a bounded sequence in $H_0^{1,B}(J,X)$. Then $\{v_n\}$ is a relatively compact sequence in $L_2(J,X)$, i.e., there exists a subsequence $\{u_m\}$ of $\{v_n\}$ such that $\{u_m\}$ is a Cauchy sequence of $L_2(J,X)$.

PROOF. Set $\tilde{v}_n = U^{-1}v_n$. Then, by (i) of Proposition 1.1, $\{v_n\}$ is a bounded sequence in $H_1(\Omega_J)$ with $\Omega_J = \{y \in \mathbb{R}^N / |y| \in J\}$. It follows from the Rellich theorem that there exists a subsequence $\{\tilde{u}_m\}$ of $\{\tilde{v}_n\}$ such that $\{\tilde{u}_m\}$ is a Cauchy sequence in $L_2(\Omega_J)$. Set $u_m = U\tilde{u}_m$. Then $\{u_m\}$ is a subsequence of $\{v_n\}$ and is a Cauchy sequence in $L_2(J,X)$.

Q.E.D.

§2. Main Theorem.

We shall now state and prove the limiting absorption principle for the operator

$$(2.1) \qquad L = -\frac{d^2}{dr^2} + B(r) + C(r)$$

which is given by (0.20) and (0.21). The potential $\mathcal{Q}(y)$ is assumed to satisfy the following

ASSUMPTION 2.1.

($\mathcal{Q}$) $\mathcal{Q}(y)$ can be decomposed as $\mathcal{Q}(y) = \mathcal{Q}_0(y) + \mathcal{Q}_1(y)$ such that $\mathcal{Q}_0$, and $\mathcal{Q}_1$ are real-valued functions on $\mathbb{R}^N$ with $N \geqq 3$.

($\mathcal{Q}_0$) $\mathcal{Q}_0 \in C^1(\mathbb{R}^N)$ and

$$(2.2) \qquad |D^j \mathcal{Q}_0(y)| \leqq c_0(1 + |y|)^{-j-\varepsilon} \qquad (y \in \mathbb{R}^N,\ j = 0, 1)$$

with constants $c_0 > 0$ and $0 < \varepsilon \leqq 1$, where D^j denotes an arbitrary derivative of j-th order.

($\mathcal{Q}_1$) $\mathcal{Q}_1 \in C^0(\mathbb{R}^N)$ and

$$(2.3) \qquad |\mathcal{Q}_1(y)| \leqq c_0(1 + |y|)^{-\varepsilon_1} \qquad (y \in \mathbb{R}^N)$$

with a constant $\varepsilon_1 > 1$ and the same constant c_0 as in ($\mathcal{Q}_0$). We set

$$(2.4) \qquad \begin{cases} C(r) = C_0(r) + C_1(r), \\ C_j(r) = \mathcal{Q}_j(r\omega)\times \qquad (j = 0, 1). \end{cases}$$

Then, by ($\mathcal{Q}_0$) and ($\mathcal{Q}_1$), we have

$$(2.5)\qquad \begin{cases} \|C_0(r)\| \leqq c_0(1+r)^{-\varepsilon}, \\ \|C_0'(r)\| \leqq c_0(1+r)^{-1-\varepsilon}, \\ \|C_1(r)\| \leqq c_0(1+r)^{-1-\varepsilon_1}, \end{cases}$$

where $\| \ \|$ denotes the operator norm of $\mathbb{B}(X)$ and $C_0'(r)$ is the strong derivative of $C_0(r)$ in $\mathbb{B}(X)$.

Let us define a class of solutions for the equation $(L-k^2)v = \ell$ with $k \in \mathbb{C}^+$ and $\ell \in F(I,X)$.

DEFINITION 2.2. (radiative function). Let δ be a fixed constant such that $\frac{1}{2} < \delta \leqq \min(\frac{1}{4}(2+\varepsilon), \frac{1}{2}\varepsilon_1)$. Let $\ell \in F(I,X)$ and $k \in \mathbb{C}^+$ be given. Then an X-valued function $v(r)$ on $\bar{I}$ is called the *radiative function* for $\{L, k, \ell\}$, if the following three conditions hold:

1) $v \in H_0^{1,B}(I,X)_{loc}$.

2) $v' - ikv \in L_{2,\delta-1}(I,X)$.

3) For all $\phi \in C_0^\infty(I,X)$ we have

$$(v, (L-\bar{k}^2)\phi)_0 = \langle \ell, \phi \rangle.$$

For the notation used above see the list of notation given at the beginning of §1. The condition 2) means $\frac{\partial v}{\partial r} - ikv$ is small at infinity, and hence 2) can be regarded as a sort of radiation condition. This is why v is called the radiative function.

THEOREM 2.3. (limiting absorption principle). Let Assumption 2.1 be fulfilled.

(i) Let $(k, \ell) \in \mathbb{C}^+ \times F(I,X)$ be given. Then the radiative function for $\{L, k, \ell\}$ is unique.

(ii) For given $(k, \ell) \in \mathbb{C}^+ \times F_\delta(I,X)$ there exists a unique radiative function $v = v(\cdot, k, \ell)$ for $\{L, k, \ell\}$ which belongs to $H^{1,B}_{0,-\delta}(I,X)$.

(iii) Let K be a compact set in $\mathbb{C}^+$. Let $v = v(\cdot, k, \ell)$ be the radiative function for $\{L, k, \ell\}$ with $k \in K$ and $\ell \in F_\delta(I,X)$. Then there exists a positive constant $C = C(K)$ depending only on K (and L) such that

(2.6) $$\|v\|_{B,-\delta} \leqq C|||\ell|||_\delta,$$

(2.7) $$\|v\|_{-\delta} + \|v'-ikv\|_{\delta-1} + \|B^{\frac{1}{2}}v\|_{\delta-1} \leqq C|||\ell|||_\delta ,$$

and

(2.8) $$\|v\|^2_{B,-\delta,(r,\infty)} \leqq C^2 r^{-(2\delta-1)}|||\ell|||^2_\delta \qquad (r \geqslant 1) .$$

(iv) The mapping: $\mathbb{C}^+ \times F_\delta(I,X) \quad (k, \ell) \to v(\cdot, k, \ell) \in H^{1,B}_{0,-\delta}(I,X)$ is continuous on $\mathbb{C}^+ \times F_\delta(I,X)$. Therefore it is also continuous as a mapping from $\mathbb{C}^+ \times F_\delta(I,X)$ into $H^{1,B}_0(I,X)_{loc}$.

Before we prove Theorem 2.3, which is our main theorem in this chapter, we prepare several lemmas, some of which will be shown in the succeeding two sections.

LEMMA 2.4. Let $k \in \mathbb{C}^+$ and let v be the radiative function for $\{L, k, 0\}$. Then v is identically zero.

The proof of this lemma will be given in §3.

For $f \in L_{2,\beta}(I,X)$ with $\beta \geqq 0$ we define $\ell[f] \in F_\beta(I,X)$ by $\langle\ell[f], \phi\rangle = (f, \phi)_0$ for $\phi \in C_0^\infty(I,X)$. It can be seen easily that

(2.9) $$|||\ell[f]|||_\beta \leqq \|f\|_\beta .$$

LEMMA 2.5. Let $k \in K$ and let v be the radiative function for $\{L, k, \ell[f]\}$ with $f \in L_{2,\delta}(I,X)$ and let $v \in L_{2,-\delta}(I,X)$, where K is a compact set in $\mathbb{C}^+$ and δ is as in Definition 2.2. Then $B^{\frac{1}{2}}v \in L_{2,\delta-1}(I,X)$ and there exists a constant $C = C(K)$ such that

$$\|v' - ikv\|_{\delta-1} + \|B^{\frac{1}{2}}v\|_{\delta-1} \leqq C\left\{\|v\|_{-\delta} + \|f\|_{\delta}\right\}, \tag{2.10}$$

and

$$\|v\|^2_{-\delta,(r,\infty)} \leqq C^2 r^{-(2\delta-1)} \left\{\|v\|^2_{-\delta} + \|f\|^2_{\delta}\right\} \quad (r \geqq 1). \tag{2.11}$$

LEMMA 2.6. Let $v_n (n = 1,2,\ldots)$ be the radiative function for $\{L, k_n, \ell_n\}$, where $k_n \in \mathbb{C}^+$ and $\ell_n \in F_\delta(I,X)$. Assume that

$$\begin{cases} k_n \to k \in \mathbb{C}+ \\ \ell_n \to \ell \qquad \text{in } F_\delta(I,X) \end{cases} \tag{2.12}$$

as $n \to \infty$, and that there exists a constant C_0 such that

$$\begin{cases} \|v_n\|_{-\delta} + \|v_n' - ik_n v_n\|_{\delta-1} \leqq C_0, \\ \|v_n\|^2_{-\delta,(r,\infty)} \leqq C_0^2 r^{-(2\delta-1)} \qquad (r \geqq 1) \end{cases} \tag{2.13}$$

for all $n = 1,2,\ldots$ Then $\{v_n\}$ has a strong limit v in $L_{2,-\delta}(I,X)$ which is the radiative function for $\{L, k, \ell\}$. Further we have

$$v_n \to v \quad \text{in } H_0^{1,B}(I,X)_{loc} \tag{2.14}$$

as $n \to \infty$.

LEMMA 2.7. Let $k_0 \in \mathbb{C}^+$ with $\mathrm{Im}k_0 > 0$ and let $\ell \in F_\beta(I,X)$ with $\beta \geqslant 0$. Then the equation

(2.15) $$(v, (L-\bar{k}_0^2)\phi)_0 = \langle \ell, \phi \rangle \quad (\phi \in C_0^\infty(I,X))$$

has a unique solution v_0 in $H_0^{1,B}(I,X)$ and v_0 satisfies

(2.16) $$\| v_0 \|_{B,\beta} \leqq C|||\ell|||_\beta \quad (C = C(k_0, \beta)) .$$

Therefore v_0 is the radiative function for $\{L, k_0, \ell\}$.

Lemmas 2.5, 2.6 and 2.7 will be proved in §4.

LEMMA 2.8. Let $v_0 \in H_0^{1,B}(I,X) \cap L_{2,\delta}(I,X)$ be a unique solution of the equation (2.15) with $\ell \in F_\delta(I,X)$ and $k_0 \in \mathbb{C}^+$, $\mathrm{Im}k_0 > 0$. Let $k \in \mathbb{C}^+$. Then the following (a) and (b) are equivalent:

(a) v is the radiative function for $\{L, k, \ell\}$.

(b) v is represented as $v = v_0 + w$, where w is the radiative function for $\{L, k, \ell[(k^2-k_0^2)v_0]\}$.

PROOF. Lemma 2.8 directly follows from the relation for $v \in L_2(I,X)_{loc}$

(2.17) $$(v - v_0, (L-\bar{k}^2)\phi)_0$$
$$= (v, (L-\bar{k}^2)\phi) - \langle \ell, \phi \rangle + ((k^2-k_0^2)v_0, \phi)_0$$
$$(\phi \in C_0^\infty(I,X))$$

and Lemma 2.7.

Q.E.D.

By the use of these lemmas Theorem 2.3 can be proved in a standard way used in the proof of the limiting absorption principle for the various operators.

PROOF of THEOREM 2.3. The proof is divided into several steps.

(I) The uniqueness of the radiative function for given $(k, \ell) \in \mathbb{C}^+ \times F(I,X)$ follows from Lemma 2.4.

(II) Let us show the estimate (2.6), (2.7) and (2.8) with $|||\ell|||_\delta$ replaced by $\|f\|_\delta$ for the radiative function v for $\{L, k, \ell[f]\}$ with $k \in K$, $v \in L_{2,-\delta}(I,X)$ and $f \in L_{2,\delta}(I,X)$. In view of Lemma 2.5 it suffices to show

$$\|v\|_{-\delta} \leqq C\|f\|_\delta , \tag{2.18}$$

because (2.6) and (2.8) with $|||\ell|||_\delta$ replaced by $\|f\|_\delta$ easily follow from (2.18) and Lemma 2.5. Let us assume that (2.18) is false. Then for each positive integer n we can find $k_n \in K$ and the radiative function v_n for $\{L, k_n\ \ell[f_n]\}$ with $f_n \in L_{2,\delta}(I,X)$ such that

$$\|v_n\|_{-\delta} = 1, \qquad \|f_n\|_\delta \leqq \frac{1}{n} \qquad (n = 1,2,\ldots) . \tag{2.19}$$

We may assume without loss of generality that $k_n \to k \in K$ as $n \to \infty$. Thus it can be seen that (2.12) and (2.13) in Lemma 2.6 hold good with $\ell_n = \ell[f_n]$, $\ell = 0$ and $C_0 = 2C$, where C is the same constant as in Lemma 2.5 and we should note that $\|\ell[f_n]\|_\delta \leqq \|f_n\|_\delta \to 0$ as $n \to \infty$. Therefore, by Lemma 2.6, $\{v_n\}$ converges in $L_{2,-\delta}(I,X)$ to the radiative function for $\{L, k, 0\}$ which satisfies $\|v\|_{-\delta} = 1$. But it follows from Lemma 2.4 that v is identically zero, which contradicts the fact that $\|v\|_{-\delta} = 1$. Thus we have shown (2.6), (2.7) (2.8) with $|||\ell|||_\delta$ replaced by $\|f\|_\delta$ for the radiative function $v \in L_{2,-\delta}(I,X)$ for $\{L, k, \ell[f]\}$ with $k \in K$ and $f \in L_{2,\delta}(I,X)$.

(III) The existence of the radiative function for (k, ℓ) with $k \in \mathbb{C}^+$, $\mathrm{Im}k > 0$ and $\ell \in F_\delta(I,X)$ has been shown by Lemma 2.7 with $\beta = \delta$. Let us next consider the case that $k \in \mathbb{R} - \{0\}$. Set $k_n = k + i/n \in \mathbb{C}^+$ and denote by v_n the radiative function for $\{L, k_n, \ell\}$. Let $k_0 \in \mathbb{C}^+$ with $\mathrm{Im}k_0 > 0$ and let v_0 be the radiative function for $\{L, k_0, \ell\}$. The existence of v_n and v_0 are assured by Lemma 2.7. Then, using (2.16) with $\beta = \delta$, we have

$$\|v_0\|_{B,\delta} \leqq C|||\ell|||_\delta \tag{2.20}$$

with $C = C(k_0)$, whence easily follows that

$$\|v_0\|^2_{-\delta,(r,\infty)} \leqq C^2 r^{-(2\delta-1)} |||\ell|||^2_\delta \qquad (r \geqq 1) . \tag{2.21}$$

On the other hand, by making use of Lemma 2.8, $w_n = v_n - v_0$ is the radiative function for $\{L, k_n, \ell[(k_n^2-k_0^2)v_0]\}$. Since v_n and v_0 belong to $L_{2,\delta}(I,X)$, $(k_n^2-k_0^2)v_0$ and $w_n = v_n - v_0$ belong to $L_{2,\delta}(I,X)$, and hence, as has been shown in (II), we have

$$\begin{cases} \|w_n\|_{-\delta} + \|w_n' - ik_n w_n\|_{\delta-1} \leqq C|k_n^2-k_0^2| \, \|v_0\|_\delta , \\ \\ \|w_n\|^2_{-\delta,(r,\infty)} \leqq C^2 r^{-(2\delta-1)} \, |k_n^2-k_0^2|^2 \, \|v_0\|^2_\delta \qquad (r \geqq 1) , \end{cases} \tag{2.22}$$

where $C = C(K_0)$ and $K_0 = \{k_n\} \cup \{k\}$. It follows (2.20), (2.21) and (2.22) that $\{k_n\}$ and $\{v_n\}$ satisfy (2.12) and (2.13) with $\ell_n = \ell$ in Lemma 2.6. Hence, by Lemma 2.6, there exists a strong limit v in $L_{2,-\delta}(I,X)$ which is the radiative function for $\{L, k, \ell\}$. Thus the existence for the radiative function for $\{L, k, \ell\}$ has been established.

(IV) Finally let us show (2.6), (2.7) and (2.8) completely. Let K be a compact set in $\mathbb{C}^+$. Let k_0 and v_0 be as in (III). Then the radiative function v for $\{L, k, \ell\}$ $(k \in K, \ell \in F_\delta(I,X))$ is represented as $v = v_0 + w$, w being the radiative function for $\{L, k, \ell[(k^2-k_0^2)v_0]\}$ by Lemma 2.8. It follows from (II) that the estimates

$$
(2.23) \qquad \begin{cases} \|w\|_{-\delta} + \|w'-ikw\|_{\delta-1} + \|B^{\frac{1}{2}}w\|_{\delta-1} \leqq C|k^2-k_0^2|\ \|v_0\|_\delta\ , \\[2ex] \|w\|^2_{-\delta,(r,\infty)} \leqq Cr^{-(2\delta-1)}|k^2-k_0^2|^2\ \|v_0\|^2_\delta \qquad (r \geqq 1) \end{cases}
$$

hold with $C = C(K)$. (2.23) is combined with (2.20) and (2.21) to give the estimates (2.7) and

$$
(2.24) \qquad \|v\|^2_{-\delta,(r,\infty)} \leqslant C^2 r^{-(2\delta-1)} |||\ell|||^2_\delta
$$

(2.6) and (2.8) directly follow from (2.7) and (2.24). The continuity of the mapping: $(k, \ell) \to v(\cdot, k, \ell)$ in $H^{1,B}_{0,-\delta}(I,X)$ is easily obtained from (III) and Lemma 2.6.

Q.E.D.

§3. Uniqueness of the Radiative Function.

This section is devoted to showing Lemma 2.4. It will be proved along the line of Jäger [1] and [2]. But some modification is necessary, since we are concerned with a long-range potential.

We shall now give a proposition related to the asymptotic behavior of the solution $v \in D(J)$, $J = (R,\infty)$ with $R \geqslant 0$, of the equation $(L - k^2)v = 0$ with $k \in \mathbf{R} - \{0\}$. Here the definition of $D(J)$ is given after the proof of Proposition 1.3 in §1.

PROPOSITION 3.1. Let Assumption 2.1 be satisfied. Let $v \in D(J)$ $(J = (R,\infty)$, $(R \geqq 0)$ satisfy the following (1) and (2):

(1) The estimate

$$|(L-k^2)v(r)|_X \leqq c(1+r)^{-2\delta}|v(r)|_X \qquad (r \in J) \tag{3.1}$$

holds with $k \in \mathbf{R} - \{0\}$ and some $c > 0$, where δ is given as in Definition 2.2.

(2) The support of v is unbounded, i.e., the set $\{r \in J/|v(r)|_X > 0\}$ is an unbounded set.

Then

$$\varliminf_{r\to\infty}\{|v'(r)|_X^2 + k^2|v(r)|_X^2\} > 0. \tag{3.2}$$

The proof will be given after proving Lemmas 3.2 and 3.3. Set for $v \in D(J)$

$$(Kv)(r) = |v'(r)|_X^2 + k^2|v(r)|_X^2 - (B(r)v(r), v(r))_X - (C_0(r)v(r), v(r)) \qquad (r \in J). \tag{3.3}$$

$Kv(r)$ is absolutely continuous on every compact interval in J.

LEMMA 3.2. Let $v \in D(J)$ and let (3.1) be satisfied. Then there exist $c_1 > 0$ and $R_1 > R$ such that

$$\frac{d}{dr}(Kv)(r) \geqq -c_1(1+r)^{-2\delta}(Kv)(r) \qquad (\text{a.e. } r \geqq R_1), \tag{3.4}$$

where a.e. means "almost everywhere".

PROOF. Setting $(L-k^2)v = f$ and using (3.1) and $(\mathcal{Q}_1)$ of Assumption 2.1, we obtain

$$\begin{aligned} \frac{d}{dr}(Kv) &= 2\mathrm{Re}\{(v'',v')_X + k^2(v,v')_X - (Bv,v')_X \\ &\quad - (C_0v,v')_X\} - \frac{d}{dr}(\{B(r) + C_0(r)\}x,x)_X|_{x=v} \\ &= 2\mathrm{Re}(C_1v-f,\ v')_X + \frac{2}{r}(Bv,v)_X - (C_0'\,v,v)_X \\ &\geqq -2c_2(1+r)^{-2\delta}|v|_X|v'|_X + \frac{2}{r}(Bv,v)_X - (C_0'v,v)_X \end{aligned} \tag{3.5}$$

with $c_2 = c + c_0$, c_0 being the same constant as in $(\mathcal{Q}_1)$ of Assumption 2.1. From the estimate

$$2|v|_X|v'|_X = \frac{\sqrt{2}}{|k|}\cdot 2\left(\frac{|k|}{\sqrt{2}}|v|_X\right)|v'|_X \leqq \frac{\sqrt{2}}{|k|}\left(|v'|_X^2 + \frac{k^2}{2}|v|_X^2\right) \tag{3.6}$$

and (3.5) it follows that we have with $c_1 = c_2\sqrt{2}/|k|$

$$\begin{aligned} \frac{d}{dr}Kv &\geqq -c_1(1+r)^{-2\delta}(|v'|_X^2 + \tfrac{1}{2}k^2|v|_X^2) + \tfrac{2}{r}(Bv,v)_X \\ &\qquad - (C_0'v,v)_X \\ &= -c_1\ (1+r)^{-2\delta}(Kv) + (\tfrac{2}{r} - c_1(1+r)^{-2\delta})\ (Bv,v)_X \\ &\quad + [\tfrac{1}{2}k^2c_1(1+r)^{-2\delta}|v|_X^2 - (\{c_1(1+r)^{-2\delta}C_0 + C_0'\}v,v)_X] \\ &\equiv -c_1(1+r)^{-2\delta}(Kv) + J_1(r) + J_2(r). \end{aligned} \tag{3.7}$$

$J_1(r) \geqq 0$ for sufficiently large r, because $2r^{-1} - c_1(1+r)^{-2\delta} \geqq 0$ for sufficiently large r. By (Q_0) of Assumption 2.1 $\|c_1(1+r)^{-2\delta}C_0(r) + C_0'(r)\| = o(r^{-2\delta})$ $(r\to\infty)$ holds, and so $J_2(r) \geqq 0$ for sufficiently large r too. Therefore there exists $R_1 > R$ such that (3.4) is valid for $r \geqq R_1$. Q.E.D.

Let us next set

$$Nv = r\{K(e^d v) + (m^2 - \log r)\, r^{-2\mu}|e^d v|_X^2\}\ , \tag{3.8}$$

where m is a positive integer, $\frac{1}{3} < \mu < \frac{1}{2}$, $d = d(r) = m(1-\mu)^{-1}r^{1-\mu}$.

<u>LEMMA 3.3.</u> Let $v \in D(J)$ satisfy (1) and (2) of Proposition 3.1. Then for fixed $\mu \in (1/3, 1/2)$ there exist $m_1 \geqq 1$ and $R_2 \geqq R$ such that

$$(Nv)(r) \geqq 0 \qquad (r \geqq R_2\ ,\ m \geqq m_1) \tag{3.9}$$

<u>PROOF.</u> Set $w = e^d v$. Then $(d/dr)(Nv)(r)$ is calculated as follows:

$$\begin{aligned} \frac{d}{dr}(Nv) &= (Kw) + r\frac{d}{dr}(Kw) + (1-2\mu)r^{-2\mu}(m^2 - \log r)\ |w|_X^2 \\ &\quad - r^{-2\mu}|w|_X^2 + 2(m^2 - \log r)r^{1-2\mu}\mathrm{Re}(w', w)_X \\ &= |w'|_X^2 + \{k^2 + (1-2\mu)r^{-2\mu}(m^2 - \log r) - r^{-2\mu}\}\ |w|_X^2 \\ &\quad - (Bw,w)_X - (C_0 w,w)_X + r\frac{d}{dr}(Kw) \\ &\quad + 2r^{1-2\mu}(m^2 - \log r)\mathrm{Re}(w',w)_X\ . \end{aligned} \tag{3.10}$$

Now we make use of the relations

$$\begin{cases} w' = e^d v' + mr^{-\mu}w \\ w'' = e^d v'' + mr^{-\mu}e^d v' + mr^{-\mu}w' - \mu m r^{-\mu-1}w \\ \quad\ = e^d v'' + 2mr^{-\mu}w' - (\mu m r^{-\mu-1} + m^2 r^{-2\mu})w \end{cases} \tag{3.11}$$

to obtain

$$(3.12)\qquad \begin{aligned}\frac{d}{dr}(Kw) &= 2\mathrm{Re}(w'' + k^2 w - C_0 w - Bw, w')_X - \frac{d}{dr}(\{B + C_0\}x,x)_X|_{x=w} \\ &= 2e^d \mathrm{Re}(C_1 v - f, w') + 4mr^{-\mu}|w'|_X^2 \\ &\quad - 2(\mu m r^{-\mu-1} + m^2 r^{-2\mu})\mathrm{Re}(w,w')_X \\ &\quad - \frac{d}{dr}(\{B + C_0\}x,x)_X|_{x=w} \quad .\end{aligned}$$

It follows from (3.10)~(3.12) that

$$(3.13)\qquad \begin{aligned}\frac{d}{dr}(Nv) &= (4mr^{1-\mu} + 1)\ |w'|_X^2 + \{k^2 + (1-2\mu)r^{-2\mu}(m^2 - \log r) - r^{-2\mu}\}|w|_X^2 \\ &\quad - (\{C_0 + rC_0'\}\ w,w)_X + 2\{re^d \mathrm{Re}(C_1 v - f, w')_X \\ &\quad - 2(\mu m r^{-\mu} + r^{1-2\mu}\log\ r)\mathrm{Re}(w,w')_X \\ &\quad + (Bw,w)_X \quad .\end{aligned}$$

By the use of (3.1) and Assumption 2.1 we arrive at

$$(3.14)\qquad \begin{aligned}\frac{d}{dr}(Nv) &\geqq [\,k^2 + m^2(1-2\mu)r^{-2\mu} - \{(1-2\mu)\ \log\ r + 1\}r^{-2\mu} \\ &\quad - 2c_0(1+r)^{-\varepsilon}]\,|w|_X^2 + (4mr^{1-\mu} + 1)\ |w'|_X^2 \\ &\quad - 2\{(c + c_0)\ (1+r)^{-(2\delta-1)} + \mu m r^{-\mu} + r^{1-2\mu}\log\ r\}|w|_X|w'|_X \quad .\end{aligned}$$

Thus we can find $R_3 > R$, independent of m, such that

$$(3.15)\qquad \begin{aligned}\frac{d}{dr}(Nv) &\geqq \{\tfrac{1}{2}k^2 + m^2(1-2\mu)r^{-2\mu}\}|w|_X^2 \\ &\quad - 2\ \{2r^{1-2\mu}\log\ r + \mu m r^{-\mu}\}|w|_X|w'|_X \\ &\quad + (4mr^{1-\mu} + 1)\ |w'|_X^2 \\ &\equiv P\ (r,m)\ |w|_X^2 - 2Q(r,m)\ |w|_X|w'|_X + S(r,m)|w'|_X^2 \\ &\qquad\qquad (r \geqq R_3, m \geqq 1)\end{aligned}$$

This is a quadratic form of $|w|_X$ and $|w'|_X$. The coefficient $P(r,m)$ of $|w|_X^2$ satisfies

(3.16) $$P \geqq k^2/2 \qquad (r>0,\ m>0) \ .$$

Further we have

$$
\begin{aligned}
PS - Q^2 &\geqq 4mr^{1-\mu}P - Q^2\\
&= 2mk^2r^{1-\mu} + 4(1-2\mu)m^3r^{1-3\mu} - \{\mu^2m^2r^{-2\mu}\\
&\quad + 4\mu mr^{1-3\mu}\log r + 4r^{2-4\mu}(\log r)^2\}\\
&= m^2\{4(1-2\mu)m - \mu^2r^{-1+\mu}\}r^{1-3\mu}\\
&\quad + m(k^2 - 4\mu r^{-2\mu}\log r)r^{1-\mu}\\
&\quad + \{mk^2 - 4r^{1-3\mu}(\log r)^2\}r^{1-\mu}\\
&\geqq k^2/2 \qquad\qquad (r\geqq R_4, m\geqq 1)
\end{aligned}
\tag{3.17}
$$

with sufficiently large $R_4 > R_3$, R_4 can be taken independent of $m \geqq 1$.

Thus we obtain

(3.18) $$\frac{d}{dr}(Nv)(r) \geqq 0 \qquad (\text{a.e. } r \geqq R_4,\ m \geqq 1) \ .$$

On the other hand $Nv(r)$ is a polynomial of order 2 with respect to m , that is, Nv can be rewritten as

(3.19) $$
\begin{aligned}
(Nv)(r) &= re^{2d}\{|mr^{-\mu}v+v'|_X^2 + k^2|v|_X^2 - (Bv,v)_X - (C_0v,v)_X\\
&\qquad + r^{-2\mu}(m^2 - \log r)\ |v|_X^2\}\\
&= re^{2d}\{2r^{-2\mu}|v|_X^2m^2 + 2r^{-\mu}\mathrm{Re}(v,v')_Xm+(Kv-r^{-2\mu}|v|_X^2\log r)\} \ .
\end{aligned}
$$

By (2) in Proposition 3.1 the support of v is unbounded, and hence there exists $R_2 \geqq R_4$ such that

(3.20) $$|v(R_2)|_X > 0 \ .$$

Then, since the coefficient of m^2 in $(Nv)(r)$ is positive when $r = R_2$, there exists $m_1 \geqq 1$ such that

$$(3.21) \qquad (Nv)(R_2) > 0 \qquad (m \geqq m_1) .$$

(3.9) is obtained from (3.18) and (3.21).

Q.E.D.

PROOF of PROPOSITION 3.1. First we consider the case that there exists a sequence $\{t_n\} \subset J$, $t_n \uparrow \infty (n \to \infty)$, such that $(Kv)(t_n) > 0$ for $n = 1, 2, \ldots$. Then there exists $r_0 \geqq R_1$ such that $(Kv)(r_0) > 0$, R_1 being as in Lemma 3.2. Let us show that $(Kv)(r) > 0$ for all $r \geqq r_0$. In fact, multiplying (3.4) by $\exp\{c_1 \int_{r_0}^{r} (1+t)^{-2\delta} dt\}$, we have

$$(3.22) \qquad \frac{d}{dr}\left\{e^{c_1\int_{r_0}^{r}(1+t)^{-2\delta}dt} \cdot (Kv)(r)\right\} \geqq 0 \qquad (\text{a.e. } r \geqq r_0),$$

whence directly follows

$$(3.23) \qquad (Kv)(r) \geqq e^{-c_1\int_{r_0}^{r}(1+r)^{-2\delta}dt}(Kv)(r_0) > 0 \qquad (r \geqq r_0) .$$

It follows from (3.23) that

$$\begin{aligned} |v'(r)|_X^2 + k^2|v(r)|_X^2 &= (Kv)(r) + (B(r)v(r), v(r))_X \\ &\quad + (C_0(r)v(r), v(r))_X \\ (3.24) \qquad &\geqq e^{-c_1\int_{r_0}^{r}(1+t)^{-2\delta}dt}(Kv)(r_0) \\ &\quad - c_0(1+r)^{-\varepsilon}k^{-2}(|v'(r)|_X^2 + k^2|v(r)|_X^2) \qquad (r \geqq r_0), \end{aligned}$$

which implies that

$$(3.25) \qquad \varliminf_{r\to\infty}(|v'(r)|_X^2 + k^2|v(r)|_X^2) \geqq e^{-\int_{r_0}^{\infty}(1+t)^{-2\delta}dt}(Kv)(r_0) > 0 \ .$$

Next consider the case that $(Kv)(r) \leqq 0$ for all $r \geqq R_5$ with some $R_5 > R_2$, R_2 being as in Lemma 3.3. Then it follows from (3.19) and (3.9) that

$$(3.26) \qquad 2r^{-2\mu}|v(r)|_X^2\, m^2 + 2r^{-\mu}\mathrm{Re}(v(r),v'(r))_X m - r^{-2\mu}|v(r)|_X^2 \log r \geqq 0 \qquad (r \geqq R_5,\ m \geqq m_1) \ ,$$

which, together with the relation

$$(3.27) \qquad \frac{d}{dr}|v(r)|_X^2 = 2\mathrm{Re}(v(r),\ v'(r))_X \ ,$$

implies that

$$(3.28) \qquad \frac{d}{dr}|v(r)|_X^2 \geqq r^{-\mu}\{\frac{1}{m_1}\log r - 2m_1\}|v(r)|_X^2 \geqq 0 \qquad (r \geqq R_6)$$

with sufficiently large $R_6 \geqq R_5$. Because of the unboundedness of the support of $v(r)$ we have $|v(R_7)|_X > 0$ with some $R_7 \geqq R_6$, and hence it can be seen from (3.28) that

$$(3.29) \qquad |v(r)|_X \geqq |v(R_7)|_X > 0 \qquad (r \geqq R_7) \ ,$$

whence follows that

$$(3.30) \qquad \varliminf_{r\to\infty}\{|v'(r)|_X^2 + k^2|v(r)|_X^2\} \geqq k^2|v(R_7)|_X^2 > 0 \ .$$

Q.E.D.

In order to show Lemma 2.4 we need one more proposition which is a corollary of Proposition 1.3 (regularity theorem).

PROPOSITION 3.4. Let $Q(y)$ be a real-valued continuous function on $\mathbb{R}^N$ and let $v \in L_2(I,X)_{loc}$ be a solution of the equation

$$(v, (L-\bar{k}^2)\phi)_0 = (f,\phi)_0 \qquad (\phi \in C_0^\infty(I,X)) \tag{3.31}$$

with $k \in \mathbb{C}^+$ and $f \in L_2(I,X)_{loc}$. Then

$$|v'(r) - ik\, v(r)|_X^2 = |v'(r) + k_2 v(r)|_X^2 + k_1^2|v(r)|_X^2 + 4k_1^2 k_2 \|v\|_{0,(0,r)}^2 + 2k_1 I_m(f,v)_{0,(0,r)} \tag{3.32}$$

holds for all $r \in I$, where $k_1 = \mathrm{Re}k$ and $k_2 = \mathrm{Im}k$.

PROOF. As has been shown in the proof of Proposition 1.3, $v \in UH_2(\mathbb{R}^N)_{loc}$ and there exists a sequence $\{\varphi_n\} \subset C_0^\infty(\mathbb{R}^N)$ such that

$$\begin{cases} \varphi_n \to \tilde{v} = U^{-1}v & \text{in } H_2(\mathbb{R}^N)_{loc}, \\ (T-k^2)\varphi_n \to \tilde{f} = U^{-1}f & \text{in } L_2(\mathbb{R}^N)_{loc} \end{cases} \tag{3.33}$$

as $n \to \infty$, where $T = -\Delta + Q(y)$. Set $\phi_n = U\varphi_n$.
Then, proceeding as in the proof of Proposition 1.3 and using the relation $(L-k^2)U = U(T-k^2)$, we have

$$\begin{cases} \phi_n \to v & \text{in } L_2(I,X)_{loc}, \\ \phi_n(r) \to v(r) & \text{in } X \quad (r \in \bar{I}), \\ \phi_n'(r) \to v'(r) & \text{in } X \quad (r \in I), \\ (L-k^2)\phi_n \to f & \text{in } L_2(I,X)_{loc} \end{cases} \tag{3.34}$$

as $n \to \infty$. Set $f_n = (L-k^2)\phi_n$, integrate $((L-k^2)\,\phi_n, \phi_n)_X = (f_n\, \phi_n)_X$ from 0 to r and make use of partial integration. Then we arrive at

$$(3.35)\qquad \begin{aligned}(\phi_n', \phi_n')_{0,(0,r)} + (\{B+C-k^2\}\phi_n, \phi_n)_{0,(0,r)} - (\phi_n'(r), \phi_n(r))_X \\ = (f_n, \phi_n)_{0,(0,r)}\ .\end{aligned}$$

By letting $n \to \infty$ in (3.35) after taking the imaginary part of it (3.35) yeilds

$$(3.36)\qquad I_m(v'(r), v(r))_X = -2k_1k_2\|v\|_{0,(0,r)} - I_m(f,v)_{0,(0,r)}.$$

(3.32) directly follows from (3.36) and

$$(3.37)\qquad \begin{aligned}|v'(r) - ikv(r)|_X^2 = |v'(r) + k_2v(r)|_X^2 + k_1^2|v(r)|_X^2 \\ -2k_1I_m(v'(r), v(r))_X\ .\end{aligned}$$

Q.E.D.

<u>PROOF of LEMMA 2.4.</u> Let v be the radiative function for $\{L,k,0\}$ with $k \in \mathbb{C}^+$. It suffices to show that $v = 0$. Let us note that we obtain from (3.32), by setting $f=0$,

$$(3.38)\qquad |v'(r) - ikv(r)|_X^2 = |v'(r) + k_2v(r)|_X^2 + k_1^2|v(r)|_X^2 + 4k_1^2k_2\|v\|_{0,(0,r)}^2\ .$$

Let us also note the relation

$$(3.39)\qquad \varliminf_{r\to\infty} |v'(r) - ikv(r)|_X^2 = 0$$

which is implied by the fact that $v' - ikv \in L_{2,\delta-1}(I,X)$. If $k_2 > 0$, then it follows from (3.38) and (3.39) that

$$(3.40) \qquad \varlimsup_{r\to\infty} \|v\|^2_{0,(0,r)} \leqq \frac{1}{4k_1^2 k_2} \varlimsup_{r\to\infty} |v'(r) - ikv(r)|_X = 0 ,$$

which means $\|v\|^2_{0,(0,\infty)} = 0$, that is, $v = 0$. Next consider the case that $k_2 = 0$, i.e., $k = k_1 \in \mathbb{R} - \{0\}$. Then from (3.38) and (3.39) we obtain

$$(3.41) \qquad \varlimsup_{r\to\infty} (|v'(r)|^2_X + k^2|v(r)|^2_X) = \varlimsup_{r\to\infty} |v'(r) - ikv(r)|_X = 0 .$$

Therefore Proposition 3.1 can be applied to show that the support of $v(r)$ is bounded in I, and hence $v = 0$ by Proposition 1.4 (the unique continuation theorem).

Q.E.D

§4. Proof of the lemmas

Now we shall show the lemmas in §2 which remain unproved. After that, some more precise properties of the mapping $(k,\ell) \to v = v(\cdot,k,\ell)$ in Theorem 2.3 will be shown.

PROOF of LEMMA 2.6. Applying Proposition 1.2 (the interior estimate) with $v = v_n$, $k = k_n$ and $\ell = \ell_n$, we have

$$\|v_n\|_{B(0,R)} \leqslant C(\|v_n\|_{0,(0,R+1)} + |||\ell_n|||_{0,(0,R+1)}) \qquad (R \in I, \quad n = 1,2,\dots) \tag{4.1}$$

with $C = C(R)$, because $\{k_n\}$ is a bounded sequence. Since $\{v_n\}$ and $\{\ell_n\}$ are bounded sequences in $L_{2,-\delta}(I,X)$ and $F_\delta(I,X)$, respectively, it can be easily seen that $\|v_n\|_{0,(0,R+1)}$ and $|||\ell_n|||_{0,(0,R+1)}$ are uniformly bounded for $n = 1,2,\dots$ with a fixed positive number R. Therefore, for fixed $R \in I$, the right-hand side of (4.1) is uniformly bounded for $n = 1,2,\dots$. Thus, Proposition 1.5 can be applied to show that there exist a subsequence $\{v_{n_m}\}$ of $\{v_n\}$ and $v \in L_2(I,X)_{loc}$ such that v_{n_m} converges to v in $L_2(I,X)_{loc}$. Moreover, the norm $\|v_{n_m}\|_{-\delta,(r,\infty)}$ tends to zero uniformly for $m = 1,2,\dots$ as $r \to \infty$ by the second relation of (2.13), and hence we arrive at

$$v_{n_m} \to v \quad \text{in} \quad L_{2,-\delta}(I,X) \tag{4.2}$$

as $m \to \infty$. By letting $m \to \infty$ in the relation

$$(v_{n_m},(L - \overline{k}_{n_m}^2)\phi) = \langle \ell_{n_m},\phi\rangle \qquad (\phi \in C_0^\infty(I,X)), \tag{4.3}$$

v is seen to be a solution of the equation

$$(v,(L= \overline{k}^2)\phi) = \langle \ell,\phi\rangle \qquad (\phi \in C_0^\infty(I,X)). \tag{4.4}$$

Since $u = v_{n_m} - v_{n_p}$ satisfies the equation

$$(4.5)\quad (u,(L-\overline{k}^2_{n_m})\Phi) = \langle \ell[(k^2_{n_m}-k^2_{n_p})v_{n_p}] + \ell_{n_m}-\ell_{n_p},\Phi\rangle \quad (\Phi \in C_0^\infty(I,X)),$$

Proposition 1.2 can be applied again to show that

$$(4.6)\quad \|v_{n_m}-v_{n_p}\|_{B,(0,R)} \leqslant C\{\|v_{n_m}-v_{n_p}\|_{0,(0,R+1)} + |k^2_{n_m}-k^2_{n_p}|\|v_{n_p}\|_{0,(0,R+1)}$$

$$+ |||\ell_{n_m}-\ell_{n_p}|||_{0,(0,R+1)}\} \to 0$$

as $m,p \to \infty$, whence follows that

$$(4.7)\qquad v_{n_m} \to v \quad \text{in} \quad H_0^{1,B}(I,X)_{loc}$$

as $m \to \infty$. Thus, letting $m \to \infty$ in the estimate

$$(4.8)\qquad \|v'_{n_m} - ik_{n_m}v_{n_m}\|_{\delta-1,(0,R)} \leqslant C_0,$$

which is obtained from the first relation of (2.13), we see that

$$(4.9)\qquad \|v' - ikv\|_{\delta-1,(0,R)} \leqslant C_0,$$

and hence, because of the arbitrariness of $R > 0$, $v'-ikv \in L_{2,\delta-1}(I,X)$. Therefore v is the radiative function for $\{L,k,\ell\}$.

As is easily seen from the discussion above, any subsequence of $\{v_n\}$ contains a subsequence which converges in $L_{2,-\delta}(I,X) \cap H_0^{1,B}(I,X)_{loc}$ to the radiative function v for $\{L,k,\ell\}$. Therefore, it follows that $\{v_n\}$ itself converges to v in $L_{2,-\delta}(I,X) \cap H_0^{1,B}(I,X)_{loc}$. *Q.E.D.*

Let us turn to the proof of Lemma 2.7. To this end, we have to investigate some properties of the function space $H^{1,B}_{0,-\beta}(I,X)$. It is a Hilbert space obtained by the completion of $C_0^\infty(I,X)$ by the norm

$$(4.10)\qquad \|\Phi\|^2_{B,-\beta} = \int_I (1+r)^{-2\beta}\{|\Phi'|^2_X + |B^{\frac{1}{2}}\Phi|^2_X + |\Phi|^2_X\}dr.$$

Obviously $H_{0,-\beta}^{1,B}(I,X) \subset H_0^{1,B}(I,X)_{loc}$ and the inner product and norm are represented as

$$(4.11) \quad \begin{cases} (u,v)_{B,-\beta} = \int_I (1+r)^{-2\beta}\{(u',v')_x + (B^{\frac{1}{2}}u-B^{\frac{1}{2}}v)_x + (u,v)_x\}dr, \\[2mm] \|u\|_{B,-\beta} = [(u,u)_{B,-\beta}]^{\frac{1}{2}}. \end{cases}$$

Let t be a positive number. Then the norm $\| \ \|_{B,-\beta,t}$ defined by

$$(4.12) \quad \|u\|_{B,-\beta,t}^2 = \int_I (t+r)^{-2\beta}\{|u'|_x^2 + |B^{\frac{1}{2}}u|_x^2 + |u|_x^2\}dr$$

is equivalent to the norm $\| \ \|_{B,-\beta}$ $(= \| \ \|_{B,-\beta,1})$. The set of all anti-linear continuous functional on $H_{0,-\beta}^{1,B}(I,X)$ is $F_\beta(I,X)$, because we can easily show that the norm $\|\!|\ell|\!\|_\beta$ of $\ell \in F_\beta(I,X)$ is equivalent to the norm

$$(4.13) \quad \|\!|\ell|\!\|{}^{\sim}_\beta = \sup\{|<\ell,\Phi>|/\Phi \in C_0^\infty(I,X), \ \|\Phi\|_{B,-\beta} = 1\}.$$

It can be easily seen, too, that the norm $\|\!|\ell|\!\|_\beta$ is equivalent to the norm defined by

$$(4.14) \quad \|\!|\ell|\!\|_{\beta,t} = \sup\{|<\ell,(t+r)^\beta\Phi>|/\Phi \in C_0^\infty(I,X), \ \|\Phi\|_B = 1\}$$

for all $t > 0$.

PROOF of LEMMA 2.7. Consider a bilinear continuous functional A_t on $H_{0,-\beta}^{1,B}(I,X) \times H_{0,-\beta}^{1,B}(I,X)$ defined by

$$(4.15) \quad A_t[u,v] = (u,v)_{B,-\beta,t} - 2\beta\int_I (t+r)^{-2\beta-1}(u,v')dr$$

$$+ \int_I (t+r)^{-2\beta}(\{C - k_0^2 - 1\}u,v)_x dr,$$

where $t > 0$ and $k_0 \in \mathbb{C}^+$ with $Im \ k_0 > 0$. Let us show that A_t is positive definite for some $t > 0$, i.e., there exists $t_0 > 0$ and

$C = C(t_0,k_0) > 0$ such that

$$(4.16)\qquad |A_{t_0}[u,u]| \geqslant C\,\|u\|^2_{B,-\beta,t_0} \qquad (u \in H^{1,B}_{0,-\beta}(I,X)).$$

It suffices to show (4.16) for $u = \Phi \in C_0^\infty(I,X)$. Set $k_0^2 = \lambda + i\mu$ $(\mu \neq 0)$ and $\tilde{C}(r) = C(r) - 1 - \lambda$. Then we have

$$(4.17)\qquad |A_t[\Phi,\Phi]|^2 = (\|\Phi\|^2_{B,-\beta,t} + \int_I (t+r)^{-2\beta}(\tilde{C}\Phi,\Phi)_X dr$$
$$- 2\beta\int_I (t+r)^{-2\beta-1}\mathrm{Re}(\Phi,\Phi')_X dr)^2$$
$$+ (\mu\int_I (t+r)^{-2\beta}|\Phi|_X^2 dr$$
$$- 2\beta\int_I (t+r)^{-2\beta}\mathrm{Im}(\Phi,\Phi')_X dr)^2.$$

By the use of a simple inequality $(a \pm b)^2 \geqslant (1-\alpha)a^2 + (1-\alpha^{-1})b^2$ $(\alpha > 0,\ a,b \in \mathbb{R})$ we have from (4.17)

$$(4.18)\qquad |A_t[\Phi,\Phi]|^2 \geqslant \tfrac{1}{2}\{(\|\Phi\|^2_{B,-\beta,t} + \int_I (t+r)^{-2\beta}(\tilde{C}\Phi,\Phi)_X dr)^2$$
$$+ \mu^2(\int_I (t+r)^{-2\beta}|\Phi|_X^2 dr)^2\}$$
$$- 4\beta^2\{(\int_I (t+r)^{-2\beta-1}\mathrm{Re}(\Phi,\Phi')_X dr)^2$$
$$+ (\int_I (t+r)^{-2\beta-1}\mathrm{Im}(\Phi,\Phi')_X dr)^2\}$$
$$\equiv \tfrac{1}{2} I_1(t) - 4\beta^2 I_2(t).$$

$I_1(t)$ can be estimated as follows:

$$(4.19)\qquad I_1(t) \geqslant (1-\alpha)\|\Phi\|^4_{B,-\beta,t} + \{\mu^2 - (\tfrac{1}{\alpha} - 1)\|\tilde{C}\|\}(\int_I (t+r)^{-2\beta}|\Phi|_X^2 dr)^2$$
$$(0 < \alpha < 1).$$

Set $\alpha = 2\|\tilde{C}\|(2\|\tilde{C}\| + \mu^2)^{-1}$ in (4.19). Then we arrive at

$$(4.20)\qquad I_1(t) \geqslant \frac{\mu^2}{2\|\tilde{C}\|+\mu^2}\|\Phi\|^4_{B,-\beta,t}.$$

On the other hand, we have

$$(4.21)\qquad I_2(t)/\|\Phi\|^4_{B,-\beta,t} \leqslant 2(\int_I (t+r)^{-2\beta}|\Phi|_X|\Phi'|_X dr)^2/t^2\|\Phi\|^2_{B,-\beta,t} \to 0$$

$$(t \to \infty).$$

It follows from (4.20) and (4.21) that A_{t_0} is positive definite with some t_0.

Let $\ell \in F_\beta(I,X)$. By the Riesz Theorem, there exists a unique $f \in H^{1,B}_{0,-\beta}(I,X)$ such that

$$(4.22)\qquad \begin{cases} \langle \ell,\Phi\rangle = (f,\Phi)_{B,-\beta,t_0}, \\ \|f\|_{B,-\beta,t_0} \leqslant C\, |||\ell|||_\beta \end{cases}$$

where $C = C(\beta,t_0)$ does not depend on f or ℓ. Now the Lax-Milgram Theorem (see, e.g., Yosida [1], p. 92) can be applied to show that there exists $w \in H^{1,B}_{0,-\beta}(I,X)$ such that

$$(4.23)\qquad \begin{cases} A_{t_0}[w,\Phi] = (f,\Phi)_{B,-\beta,t_0} & (\Phi \in C_0^\infty(I,X)) \\ \|w\|_{B,-\beta,t_0} \leqslant C\|f\|_{B,-\beta,t_0} & (C = C(\beta,t_0)). \end{cases}$$

By partial integration, we obtain, from (4.22) and (4.23),

$$(4.24)\qquad ((t_0 + r)^{-2\beta}w,\ (L - \overline{k}_0^2)\Phi)_0 = \langle \ell,\Phi\rangle \quad (\Phi \in C_0^\infty(I,X)).$$

(4.22) and (4.23) can be used again to see that $v = (t_0 + r)^{-2\beta}w$ satisfies the estimate (2.16). Thus, $v = (t_0 + r)^{-2\beta}w$ is the radiative function for $\{L,k_0,\ell\}$. The uniqueness of the radiative function has been already proved by Lemma 2.4. *Q.E.D.*

In order to show Lemma 2.5, we have to prepare two lemmas.

LEMMA 4.1. Let K be a compact set in $\mathbb{C}^+$. Then there exists a positive constant $C = C(K)$ such that

$$k_2\|v\|_{1-\delta} \leqslant C\{\|v\|_{-\delta} + \|v' - ikv\|_{\delta-1} + \|f\|_{\delta}\} \tag{4.25}$$

holds for any radiative function v for $\{L,k,\ell[f]\}$ with $k = k_1 + ik_2 \in K$, $k_2 > 0$ and $f \in L_{2,\delta}(I,X)$.

Proof. Let $k \in K$ with $\operatorname{Im} k = k_2 > 0$ and let $f \in L_{2,\delta}(I,X)$. Then, by Lemma 2.7, there exists a unique radiative function $v = v(\cdot,k,\ell[f])$ such that $v,v' \in L_{2,\delta}(I,X)$. Moreover, $v \in D(I) \cap H_0^{1,B}(I,X)_{loc}$ by Proposition 1.3 and v satisfies the equation $(L - k^2)v = f$ for almost all $r \in I$. Multiply the both sides of $((L-k^2)v(r),v(r))_X = (f(r),v(r))_X$ by $(1+r)^{2-2\delta}$, integrate from R to T $(0 < R < T < \infty)$ and take the imaginary part. Then

$$\begin{aligned}(2-2\delta)\operatorname{Im}\int_R^T (1+r)^{1-2\delta}(v',v)_X dr - (1+T)^{2-2\delta}\operatorname{Im}(v'(T),v(T))_X \\ + (1+R)^{2-2\delta}\operatorname{Im}(v'(R),v(R))_X - 2k_1k_2\int_R^T (1+r)^{2-2\delta}|v|_X^2 dr \\ = \operatorname{Im}\int_R^T (1+r)^{2-2\delta}(f,v)_X dr.\end{aligned} \tag{4.26}$$

Since $v,v' \in L_{2,\delta}(I,X)$, $\underline{\lim}_{r\to\infty}(1+r)^{2-2\delta}(v'(r),v(r))_X = 0$ holds, and $\lim_{r\to 0}\operatorname{Im}(v'(r),v(r)) = 0$ follows from (3.36). Therefore, letting $T \to \infty$ along a suitable sequence $\{T_n\}$ and $R \to 0$, we have

$$\begin{aligned}k_2\|v\|_{1-\delta}^2 &\leqslant \frac{1}{2|k_1|}\{(2-2\delta)\int_I(1+r)^{1-2\delta}|v'|_X|v|_X dr + \int_I(1+r)^{2-2\delta}|f|_X|v|_X dr\} \\ &\equiv \frac{1}{2|k_1|}\{(2-2\delta)J_1 + J_2\}.\end{aligned} \tag{4.27}$$

Let us estimate J_1 and J_2.

$$(4.28) \quad \begin{cases} J_1 \leqslant \| v' \|_{-\delta} \| v \|_{1-\delta} \leqslant \{ \| v'-ikv \|_{-\delta} + a \| v \|_{-\delta} \} \| v \|_{1-\delta} \quad (a = \max_{k \in K} |k|), \\[2ex] J_2 \leqslant \| f \|_{1-\delta} \| v \|_{1-\delta}, \end{cases}$$

where we used the Schwarz inequality. Set $b = \min_{k \in K} |k_1|$. Then it follows from (4.27) and (4.28) that

$$(4.29) \quad \begin{aligned} k_2 \| v \|_{1-\delta} &\leqslant \frac{1}{2b} \left(\| v'-ikv \|_{-\delta} + a \| v \|_{-\delta} + \| f \|_{1-\delta} \right) \\[2ex] &\leqslant \frac{1}{2b} \left(\| v'-ikv \|_{\delta-1} + a \| v \|_{-\delta} + \| f \|_{\delta} \right), \end{aligned}$$

where it should be noted that $-\delta < \delta-1$ and $1-\delta < \delta$. (4.25) follows directly from (4.29). Q.E.D.

LEMMA 4.2. Let $v \in D(T)$ and $k \in \mathbb{C}^+$. Let $\xi(r)$ be a real-valued C^2 function on I such that $0 \leqslant \xi \leqslant 1$, and

$$(4.30) \quad \xi(r) = \begin{cases} 0 & (r \leqslant R) \\[2ex] 1 & (r \leqslant R+1) \end{cases}$$

with $R > 0$.

$$(4.31) \quad \begin{aligned} \frac{\alpha}{2} \int_R^T &\xi(1+r)^{\alpha-1} |v'-ikv|_x^2 dr + \left(1 - \frac{\alpha}{2}\right) \int_R^T \xi(1+r)^{\alpha-1} |B^{\frac{1}{2}}v|_x^2 dr \\[2ex] &\leqslant \mathrm{Re} \int_R^T \xi(1+r)^{\alpha}(f, v'-ikv)_x dr - \mathrm{Re} \int_R^T \xi(1+r)^{\alpha}(C_1 v, v'-ikv)_x dr \\[2ex] &\quad + \frac{1}{2} \int_R^T \xi(1+r)^{\alpha}(C_0' v, v)_x dr + \frac{\alpha}{2} \int_R^T \xi(1+r)^{\alpha-1}(C_0 v, v)_x dr \\[2ex] &\quad - k_2 \int_R^T \xi(1+r)^{\alpha}(C_0 v, v)_x dr \\[2ex] &\quad + \frac{1}{2} \int_R^T \xi'(1+r)^{\alpha}\{ |B^{\frac{1}{2}}v|_x^2 + (C_0 v, v)_x \} dr \\[2ex] &\quad + \frac{1}{2} (1+T)^{\alpha}\{ |v'(T)-ikv(T)|_x^2 - (C_0(T)v(T), v(T))_x \}, \end{aligned}$$

where $f = (L-k^2)v$, $\alpha \in \mathbb{R}$ and $T \geqslant R+1$.

PROOF. The relation $(L-k^2)v = f$ can be rewritten as

$$(4.32) \qquad -(v'-ikv)' - ik(v'-ikv) + (B(r) + C_0(r) + C_1(r))v = f$$

whence follows that

$$(4.33) \qquad -((v'-ikv),v'-ikv)_X - ik|v'-ikv|_X^2 + ((B + C_0 + C_1)v,v'-ikv)_X = (f,v'-ikv)_X.$$

Take the real part of (4.33) and note that $k_2 \geqslant 0$. Then we obtain

$$(4.34) \qquad \begin{aligned} &-\frac{1}{2}\frac{d}{dr}\{|v'-ikv|_X^2\} + \frac{1}{2}\frac{d}{dr}\{|B^{\frac{1}{2}}v|_X^2\} - \frac{1}{2}\left(\left\{\frac{dB(r)}{dr}\right\}v,v\right)_X \\ &+ k_2(C_0v,v)_X - \frac{1}{2}\frac{d}{dr}\{(C_0v,v)_X\} - \frac{1}{2}(C_0'v,v)_X \\ &+ \mathrm{Re}(C_1v,v'-ikv)_X \leqslant \mathrm{Re}(f,v'-ikv)_X. \end{aligned}$$

Multiplying both sides of (4.34) by $\xi(1+r)^\alpha$, integrating from R to T and making use of partial integration, we arrive at (4.31). *Q.E.D.*

PROOF of (2.10) of LEMMA 2.5. Let $v \in L_{2,-\delta}(I,X)$ be the radiative function for $\{L,k,\ell[f]\}$ with $k \in K$ and $f \in L_{2,\delta}(I,X)$. Then $v \in D(I) \cap H_0^{1,B}(I,X)_{loc}$ by Proposition 1.3 and we have $(L-k^2)v(r) = f(r)$ for almost all $r \in I$. Let $\alpha = 2\delta-1$ and $R = 1$ in Lemma 4.2. Then, it follows from (4.31) that

$$(4.35) \qquad \begin{aligned} &(\delta - \tfrac{1}{2})\|\xi^{\frac{1}{2}}(v'-ikv)\|_{\delta-1,(1,T)}^2 + (\tfrac{3}{2} - \delta)\|\xi^{\frac{1}{2}}B^{\frac{1}{2}}v\|_{\delta-1(1,T)}^2 \\ &\leqslant \int_1^T \xi(1+r)^{2\delta-1}|f|_X|v'-ikv|_X dr + c_0\int_1^T \xi(1+r)^{-2\delta}|v|_X|v'-ikv|_X dr \\ &+ \frac{c_0}{2}\int_1^T \xi(1+r)^{-2\delta}|v|_X^2 dr + \frac{2\delta-1}{2}c_0\int_1^T \xi(1+r)^{-2\delta}|v|_X^2 dr \\ &+ c_0k_2\int_1^T \xi(1+r)^{1-2\delta}|v|_X^2 dr + \frac{1}{2}\int_1^2 |\xi'|(1+r)^{2\delta-1}\{|B^{\frac{1}{2}}v|_X^2 + c_0|v|_X^2\}dr \end{aligned}$$

$$+\frac{1}{2}\{(1+T)^{2\delta-1}|v'(T)-ikv(T)|_X^2+c_0(1+T)^{1-2\delta}|v(T)|_X^2\},$$

where c_0 is the constant given in Assumption 2.1 and the following estimates have been used:

$$(4.36)\quad\begin{cases}(1+r)^{2\delta-1}\|C_1(r)\|\leqslant c_0(1+r)^{2\delta-1-\varepsilon_1}\leqslant c_0(1+r)^{-1}=c_0(1+r)^{-\delta+(\delta-1)},\\(1+r)^{2\delta-1}\|C_0'(r)\|\leqslant c_0(1+r)^{2\delta-2-\varepsilon}\leqslant c_0(1+r)^{-2\delta},\\(1+r)^{2\delta-2}\|C_0(r)\|\leqslant c_0(1+r)^{2\delta-2-\varepsilon}\leqslant c_0(1+r)^{-2\delta},\\(1+r)^{2\delta-1}\|C_0(r)\|\leqslant c_0(1+r)^{2\delta-1-\varepsilon}\leqslant c_0(1+r)^{1-2\delta}.\end{cases}$$

By Lemma 4.1 and the Schwarz inequality, we have

the right-hand side of (4.35)

$$(4.37)\quad\begin{aligned}&\leqslant\|f\|_\delta\|\xi^{\frac12}(v'-ikv)\|_{\delta-1}+c_0\|v\|_{-\delta}\|\xi^{\frac12}(v'-ikv)\|_{\delta-1}\\&\quad+\delta c_0\|v\|_{-\delta}^2+c_0C\|v\|_{-\delta}\{\|v\|_{-\delta}+\|v'-ikv\|_{\delta-1}+\|f\|_\delta\}\\&\quad+\frac{1}{2}(1+c_0)C_\xi 3^{2\delta-1}\|v\|_{B,(1,2)}^2\\&\quad+\frac{1}{2}(1+T)\{(1+T)^{2\delta-2}|v'(T)-ikv(T)|_X^2+c_0(1+T)^{-2\delta}|v(T)|_X^2\}\end{aligned}$$

with $C_\xi=\max_r|\xi'(r)|$. Since $v'-ikv\in L_{2,\delta-1}(I,X)$ and $v\in L_{2,-\delta}(I,X)$, the last term of (4.37) vanishes as $T\to\infty$ along a suitable sequence $\{T_n\}$, and hence we obtain from (4.35) and (4.37)

$$(4.38)\quad\begin{aligned}&\|v'-ikv\|_{\delta-1,(2,\infty)}+\|B^{\frac12}v\|_{\delta-1,(2,\infty)}\\&\leqslant C\{\|v\|_{-\delta}+\|f\|_\delta+\|v'-ikv\|_{0,(1,2)}+\|v\|_{B,(1,2)}\}.\end{aligned}$$

(2.10) follows from (4.38) and Proposition 1.2 (the interior estimate).

Q.E.D.

PROOF of (2.11) of LEMMA 2.5. From (3.32) in Proposition 3.4, we have

$$(4.39)\qquad |v(t)|_X^2 \leqslant k_1^{-2}|v'(t)-ikv(t)|_X^2 + 2|k_1|^{-1}\|f\|_\delta\|v\|_{-\delta}.$$

Multiply (4.39) by $(1+t)^{-2\delta}$ and integrate from r to ∞. Then, it follows that

$$\|v\|^2_{-\delta,(\gamma,\infty)} \leqslant \frac{1}{k_1^2}\int_r^\infty (1+t)^{-2\delta}|v'-ikv|_X^2 dt + \frac{2}{|k_1|(2\delta-1)}(1+r)^{-(2\delta-1)}\|f\|_\delta\|v\|_{-\delta}$$

$$(4.40)\qquad \leqslant \frac{1}{k_1^2}(1+r)^{-2(2\delta-1)}\|v'-ikv\|^2_{\delta-1,(r,\infty)} + \frac{2}{|k_1|(2\delta-1)}(1+r)^{-(2\delta-1)}\|f\|_\delta\|v\|_{-\delta}.$$

(2.11) is obtained from (4.40) and (2.10). *Q.E.D.*

Now that all the lemmas in §2 and Theorem 2.3 have been proved completely, we can show more precise properties of the mapping

$$(4.41)\qquad \mathbb{C}^+ \times L_{2,}(I,X) \ni (k,\ell) \to v(\cdot,k,\ell[f]) \in L_{2,-\delta}(I,X)$$

by reexamining the proof of the lemmas in §2.

LEMMA 4.3. Let $\{f_n\}$ be a sequence in $L_{2,\delta}(I,X)$ such that f_n converges weakly to $f \in L_{2,\delta}(I,X)$ as $n \to \infty$. Let $\{k_n\}$ be a sequence in $\mathbb{C}^+$ such that $k_n \to k \in \mathbb{C}^+$ with $k \in \mathbb{C}^+$ as $n \to \infty$. Let v_n be the radiative function for $\{L,k_n,\ell[f_n]\}$. Then, there exists a subsequence $\{v_{n_m}\}$ of $\{v_n\}$ such that

$$(4.42)\qquad v_{n_m} \to v \quad \text{in} \quad L_{2,-\delta}(I,X) \cap H_0^{1,B}(I,X)_{loc}$$

as $m \to \infty$, where v denotes the radiative function for $\{L,k,\ell[f]\}$.

PROOF. Since $\{f_n\}$ is weakly convergent, $\{f_n\}$ is a bounded sequence in $L_{2,\delta}(I,X)$. Therefore, it follows from (2.7) in Theorem 2.3 that $\{v_n\}$ is a bounded sequence in $L_2(I,X)_{loc}$, that is, $\{\tilde{v}_n\}(\tilde{v}_n = U^{-1}v_n)$ is a bounded sequence in $L_2(\mathbb{R}^N)_{loc}$. Noting that $\tilde{v}_n$ satisfies the equation

$$(4.43) \qquad -\Delta\tilde{v}_n + Q\tilde{v}_n - k_n^2\tilde{v}_n = \tilde{f}_n \quad (\tilde{f}_n = U^{-1}f_n),$$

we can see that $\{v_n\}$ is really a bounded sequence in $H_2(\mathbb{R}^N)_{loc}$. Then, by the repeated use of the Rellich Theorem, it can be shown that there exists a subsequence $\{\tilde{v}_{n_m}\}$ of $\{\tilde{v}_n\}$ which is a Cauchy sequence in $H_1(\mathbb{R}^N)_{loc}$. Therefore, $\{v_{m_n}\}$ is a Cauchy seuqence in $H_0^{1,B}(I,X)_{loc}$ with the limit $v \in H_0^{1,B}(I,X)_{loc}$ r, it follows from (2.8) in Theorem 2.3 that $\{v_{n_m}\}$ converges $L_{2,-\delta}(I,X)$, too. In quite a similar way to the one used in . of Lemma 2.6, we can easily show that v is the radiative function for $\{L,k,\ell[f]\}$. *Q.E.D.*

From the above lemma, we obtain

THEOREM 4.4. Let Assumption 2.1 be satisfied.

(i) Then the mapping

$$(4.44) \qquad \mathbb{C}^+ \ni k \to v(\cdot,k,\ell[\cdot]) \in L_{2,-\delta}(I,X)$$

is a $\mathbb{B}(L_{2,\delta}(I,X),L_{2,-\delta}(I,X))$-valued continuous function on $\mathbb{C}^+$, that is, if we set

$$(4.45) \qquad v(\cdot,k,\ell[f]) = (L - k^2)^{-1}f \quad (f \in L_{2,\delta}(I,X)),$$

then $(L - k^2)^{-1} \in \mathbb{B}(L_{2,\delta}(I,X),L_{2,-\delta}(I,X))$ and $(L - k^2)^{-1}$ is continuous in $k \in \mathbb{C}^+$ in the sense of the operator norm of $\mathbb{B}(L_{2,\delta}(I,X),L_{2,-\delta}(I,X))$.

(ii) For each $k \in \mathbb{C}^+$, $(L - k^2)^{-1}$, defined above, is a compact operator from $L_{2,\delta}(I,X)$ into $L_{2,-\delta}(I,X)$.

PROOF. Suppose that the assertion (i) is false at a point $k \in \mathbb{C}^+$. Then, there exist a positive number $\beta > 0$ and sequences $\{f_n\} \subset L_{2,\delta}(I,X)$ $\{k_n\} \subset \mathbb{C}^+$ such that

$$(4.46) \qquad \begin{cases} \|f_n\|_\delta = 1 & (n = 1,2,\ldots), \\ k_n \to k & (n \to \infty), \\ \|v(\cdot,k,\ell[f_n]) - v(\cdot,k_n,\ell[f_n])\| \geqslant \beta & (n = 1,2,\ldots). \end{cases}$$

With no loss of generality, we may assume that f_n converges weakly in $L_{2,\delta}(I,X)$ to f with $f \in L_{2,\delta}(I,X)$. Then, Lemma 4.3 can be applied to show that there exists a subsequence $\{n_m\}$ of positive integers such that

$$(4.47) \qquad \begin{cases} v(\cdot,k_{n_m},\ell[f_{n_m}]) \to v(\cdot,k,\ell[f]), \\ \\ v(\cdot,k,\ell[f_{n_m}]) \to v(\cdot,k,\ell[f]) \end{cases}$$

in $L_{2,-\delta}(I,X)$ as $m \to \infty$. Therefore, we obtain

$$(4.48) \qquad \|v(\cdot,k,\ell[f_{n_m}]) - v(\cdot,k_{n_m},\ell[f_{n_m}])\|_{-\delta} \to 0$$

as $m \to \infty$, which contradicts the third relation of (4.46). Thus, the proof of (i) is complete. (ii) follows directly from Lemma 4.3. *Q.E.D.*

Finally, we shall prove a theorem which shows continuous dependence of the radiative function on the operator $C(r)$. This will be useful in §6.

THEOREM 4.5. Let L_n, $n = 1,2,\ldots$, be the operators of the form

$$(4.49) \qquad L_n = -\frac{d^2}{dr^2} + B(r) + C_n(r), \quad C_n(r) = C_{0n}(r) + C_{01}(r) \quad (r \in I)$$

with $C_{jn}(r) = \mathcal{Q}_{jn}(r\omega)\times$ for $j = 0,1$. Here $\mathcal{Q}_{0n}(y)$ and $\mathcal{Q}_{1n}(y)$ are assumed to be real-valued functions on $\mathbb{R}^N$ which satisfy (Q_0) and (Q_1) of Assumption 2.1 with Q_0 and Q_1 replaced by Q_{0n} and Q_{1n}, respectively. The constants $\varepsilon, \varepsilon_1$ and c_0 in (Q_0) and (Q_1) are assumed to be independent of $n = 1,2,\ldots$. Further assume that

$$\lim_{n\to\infty} Q_{jn}(y) = Q_j(y) \quad (j = 0,1,\ y \in \mathbb{R}^N) \tag{4.50}$$

with $Q_0(y)$ and $Q_1(y)$ which satisfy Assumption 2.1. Let v_n, $n = 1,2,\ldots$, be the radiative function for $\{L_n, k_n, \ell_n\}$ such that $k_n \in \mathbb{C}^+$, $\ell_n \in F_\delta(I,X)$ and

$$\begin{cases} k_n \to k & \text{in } \mathbb{C}^+ \\ \ell_n \to \ell & \text{in } F_\delta(I,X) \end{cases} \tag{4.51}$$

as $n \to \infty$ with $k \in \mathbb{C}^+$ and $\ell \in F_\delta(I,X)$. Then we have

$$v_n \to v \quad \text{in } H^{1,B}_{0,-\delta}(I,X), \tag{4.52}$$

where v is the radiative function for $\{L,k,\ell\}$. And there exists a constant C such that

$$\begin{cases} \|v_n\|_{B,-\delta} \leqslant C\, |||\ell|||_\delta, \\ \|v_n' - ik_n v_n\|_{\delta-1} + \|B^{\frac{1}{2}} v_n\|_{\delta-1} \leqslant C\, |||\ell|||_\delta, \\ \|v_n\|^2_{B,-\delta,(r,\infty)} \leqslant C^2 r^{-(2\delta-1)}\, |||\ell|||_\delta \quad (r \geqslant 1) \end{cases} \tag{4,53}$$

for all $n = 1,2,\ldots$. $C = C(k)$ is bounded when k moves in a compact set in $\mathbb{C}^+$.

PROOF. Let g_n be the radiative function for $\{L_n, k_0, \ell_n\}$ and let w_n be the radiative function for $\{L_n, k_n \ell[(k_n^2 - k_0^2) g_n]\}$, where $k_0 \in \mathbb{C}^+$, $\operatorname{Im} k_0 > 0$. We have $v_n = g_n + w_n$ by Lemma 2.8. Reexamining the proof of Lemmas 2.5 and 2.7, we can find a constant C, which is independent of

$n = 1,2,\ldots,$ such that

$$(4.54)\qquad \begin{cases} \|g_n'\|_\delta + \|B^{\frac{1}{2}}g_n\|_\delta + \|g_n\|_\delta \leqslant C\,|||\ell_n|||_\delta, \\ \|w_n' - ik_n w_n\|_{\delta-1} + \|B^{\frac{1}{2}}w_n\|_{\delta-1} \leqslant C(\|g_n\|_\delta + \|w_n\|_{-\delta}), \\ \|w_n\|^2_{-\delta,(r,\infty)} \leqslant C^2 r^{-(2\delta-1)}(\|g_n\|^2_\delta + \|w_n\|^2_{-\delta}) \quad (r \geqslant 1) \end{cases}$$

for all $n = 1,2,\ldots.$ Thus, we obtain, for all n

$$(4.55)\qquad \begin{cases} \|v_n' - ik_n v_n\|_{\delta-1} + \|B^{\frac{1}{2}}v_n\|_{\delta-1} \leqslant C(\,|||\ell|||_\delta + \|w_n\|_{-\delta}), \\ \|v_n\|^2_{-\delta,(r,\infty)} \leqslant C^2 r^{-(2\delta-1)}(\,|||\ell|||^2_\delta + \|w_n\|^2_{-\delta}) \quad (r \geqslant 1) \end{cases}$$

with a constant $C > 0$. The estimate

$$(4.56)\qquad \|w_n\|_{-\delta} \leqslant C\|g_n\|_\delta \quad (n = 1,2,\ldots)$$

can be shown in quite a similar way to the one used in the proof of Theorem 2.3 and Lemma 2.6. In fact, if we assume to the contrary, then there is a subsequence $\{h_m\}$ of $\{w_n/\|w_n\|_{-\delta}\}$ which converges to the radiative function for $\{L,k,0\}$, where we have used the interior estimate (Proposition 1.3), (4.54) and (4.50). We have $\|h\|_{-\delta} = 1$. On the other hand, $h = 0$ by the uniqueness of the radiative function (Lemma 2.4), which is a contradiction. (4.53) follows from (4.54) and (4.56). Proceeding as in the proof of Lemma 2.6, we can show the convergence of $\{v_n\}$ to v in $L_{2,-\delta}(I,X) \cap H_0^{1,B}(I,X)_{loc}$, which, together with (4.53), implies that $\{v_n\}$ converges to v in $H^{1,B}_{0,-\delta}(I,X)$. *Q.E.D.*

Chapter II. Asymptotic Behavior of the Radiative Function

§5. Construction of a Stationary Modifier

Throughout this chapter, the potential $Q(y)$ is assumed to satisfy the following conditions, which are stronger than in the previous chapter.

ASSUMPTION 5.1.

(Q) $Q(y)$ can be decomposed as $Q(y) = Q_0(y) + Q_1(y)$ such that Q_0 and Q_1 are real-valued function on $\mathbb{R}^N$, N being an integer with $N \geqq 3$.

$(\tilde{Q}_0)$ There exist constants $c_0 > 0$, $0 < \varepsilon \leqq 1$ such that $Q_0(y)$ is a C^{m_0} function and

(5.1) $$|D^j Q_0(y)| \leqq c_0(1 + |y|)^{-j-\varepsilon} \qquad (y \in \mathbb{R}^N, \ j = 0,1,2,\dots m_0)$$

where D^j denotes arbitrary derivatives of jth order and m_0 is the least integer satisfying

(5.2) $$m_0 > \frac{2}{\varepsilon} - 1 \quad \text{and} \quad m_0 \geqq 3.$$

Let m_1 be the least integer such that $m_1\varepsilon > 1$. When $m_0 > m_1$, we assume that $(m_1 - 1)\varepsilon < 1$. Further $Q_0(y)$ is assumed to satisfy

(5.3) $$Q_0(y) \equiv 0 \qquad (|y| \leqq 1).$$

$(\tilde{Q}_1)$ $\tilde{Q}_1(y)$ is a continuous function on $\mathbb{R}^N$ and there exists a constant $\varepsilon_1 > \max(2 - \varepsilon, 3/2)$ such that

(5.4) $$|Q_1(y)| \leqq c_0(1 + |y|)^{-\varepsilon_1} \qquad (y \in \mathbb{R}^N)$$

with the same constant c_0 as in $(\tilde{Q}_0)$.

REMARK 5.2. (1) Here and in the sequel, the constant δ in Definition 2.1 is assumed to satisfy the additional conditions

$$(5.5)\qquad \begin{cases} \delta \leqq \min(\varepsilon, \varepsilon_1 - 1) & \text{when } \tfrac{1}{2} < \varepsilon \leqq 1, \\ \delta \leqq \min\left(\frac{\varepsilon}{4}(m_0+1),\ \frac{\varepsilon_1 - 1 + \varepsilon}{2}\right) & \text{when } 0 < \varepsilon \leqq \tfrac{1}{2}. \end{cases}$$

From the conditions (5.2) and (5.4), it is easy to see that (5.5) does not contradict the condition $\delta > \frac{1}{2}$.

(2) If $\varepsilon > \frac{1}{2}$, then $m_0 = 3$ and $m_1 = 2$.

(3) The condition $(m_1 - 1)\varepsilon < 1$ is trivial. In fact, when $\frac{1}{\varepsilon}$ is an integer, we can exchange ε for a little smaller and irrational ε'. The condition (5.3) is also trivial, because $\mathcal{Q}_0(y)$ and $\mathcal{Q}_1(y)$ can be replaced by $\phi(y)\mathcal{Q}_0(y)$ and $(1-\phi(y))\mathcal{Q}_0(y) + \mathcal{Q}_1(y)$, where ϕ is a real-valued C^∞ function on $\mathbb{R}^N$ such that $\phi(y) = 0$ $(|y| \leqq 1)$, $= 1 (|y| \geqq 2)$.

(4) A general short-range potential

$$(5.6)\qquad \mathcal{Q}_1(y) = O(|y|^{-1-\varepsilon_0}) \qquad (\varepsilon_0 > 0,\ |y| \to \infty)$$

does not satisfy (5.4) in Assumption 5.1. In §12, we shall discuss the Schrödinger operator with a general short-range potential.

The following is the main result of this chapter.

<u>THEOREM 5.3</u>. (asymptotic behavior of the radiative function). Let Assumption 5.1 be satisfied. Then, there exists a real-valued function $Z(y) = Z(y,k)$ on $\mathbb{R}^N \times (\mathbb{R} - \{0\})$ such that $Z \in C^3(\mathbb{R}^N)$ as a function of y and there exists the limit

$$(5.7)\qquad F(k,\ell) = \operatorname*{s-lim}_{r\to\infty} e^{-i\mu(r\cdot,k)} v(r) \quad \text{in } X$$

for any radiative function v for $\{L,k,\ell\}$ with $k \in \mathbb{R} - \{0\}$ and $\ell \in F_{1+\beta}(I,X)$, where $\mu(y,k)$ is defined by

$$(5.8)\qquad \begin{cases} \mu(y,k) = rk - \lambda(y,k) \\ \\ \lambda(y,k) = \int_0^r Z(t\omega,k)dt \end{cases}$$

with $r = |y|$, $\omega = y/|y|$ and

$$(5.9)\qquad \beta = \begin{cases} \delta-\varepsilon & \text{if } 0 < \varepsilon \leq \frac{1}{2} \\ \\ 0 & \text{if } \frac{1}{2} < \varepsilon < 1. \end{cases}$$

This theorem will be proved in §7 by making use of the next.

THEOREM 5.4. Let Assumption 5.1 be satisfied. Let v be the radiative function for $\{L,k,\ell\}$ with $k \in \mathbb{R} - \{0\}$ and $\ell \in F_{1+\beta}(I,X)$, β being as in (5.9). Then we have $u' - iku$, $B^{\frac{1}{2}}u \in L_{2,\beta}(I,X)$, where $u = e^{i\lambda}v$ and λ is given by (5.8). Further, let K be a compact set in $\mathbb{R} - \{0\}$. Then, there exists $C = C(K)$ such that

$$(5.10)\qquad \|u'-iku\|_\beta + \|B^{\frac{1}{2}}u\|_\beta \leq C|||\ell|||_{1+\beta} \qquad (u = e^{i\lambda}v)$$

and

$$(5.11)\qquad |v(r)|_X \leq C|||\ell|||_{1+\beta} \qquad (r \in \bar{I})$$

for any radiative function v for $\{L,k,\ell\}$ with $k \in K$ and $\ell \in F_{1+\beta}(I,X)$, where β is as in (5.9).

As was proved in Jäger [3], we may take $Z(y) \equiv 0$ when $\mathcal{Q}_0(y) \equiv 0$. Saitō [3] (and Ikebe [2]) showed that we may take

$$(5.12)\qquad Z(y) = \frac{1}{2k}\mathcal{Q}_0(y)$$

in the case that $\frac{1}{2} < \varepsilon \leq 1$. It will be shown that (5.12) is the "first approximation" of $Z(y)$ in the general case. The function $\lambda(y,k)$ is called a <u>stationary modifier</u>.

In the remainder of this section, we shall construct the kernel $Z(y,k)$ of the stationary modifier $\lambda(y,k)$. To this end, let us consider the following problem: Find a real-valued function $\lambda(y,k)$ on $\mathbb{R}^N \times (\mathbb{R} - \{0\})$ such that

$$(5.13) \qquad |(L-k^2)(e^{i\mu}x)|_X = O(r^{-\tilde{\varepsilon}}) \quad (\tilde{\varepsilon} > 1, \quad r \to \infty)$$

for any $x \in D$, where μ is given by (5.8). If $\mathcal{Q}_0(y) \equiv 0$, then $\lambda(y) \equiv 0$ is a solution of this problem. In order to solve this problem, we have to investigate some properties of the Laplace-Beltrami operator Λ_N on S^{N-1}. Let us introduce polar coordinates $(r,\theta_1,\theta_2,\ldots,\theta_{N-1})$ as in (0.22), i.e.,

$$(5.14) \qquad \begin{cases} y_1 = r\cos\theta_1, \\ y_j = r\sin\theta_1\sin\theta_2\ldots\sin\theta_{j-1}\cos\theta_j \quad (j = 2,3,\ldots,N-1), \\ y_N = r\sin\theta_1\sin\theta_2\ldots\sin\theta_{N-2}\sin\theta_{N-1}, \end{cases}$$

where $r \geqq 0$, $0 \leqq \theta_1,\theta_2,\ldots,\theta_{N-2} \leqq \pi$, $0 \leqq \theta_{N-1} < 2\pi$. We set

$$(5.15) \qquad \begin{cases} b_1 = 1, \\ b_j = b_j(\theta) = \sin\theta_1\sin\theta_2\ldots\sin\theta_{j-1} \quad (j = 2,3,\ldots,N-1), \\ M_j = M_j(\theta) = b_j(\theta)^{-1}\dfrac{\partial}{\partial\theta_j} \quad (j = 1,2,\ldots,N-1). \end{cases}$$

Then we obtain from (0.23)

$$(5.16) \qquad \Lambda_N x = \sum_{j=1}^{N-1} b_j(\theta)^{-2}(\sin\theta_j)^{-N+j+1}\frac{\partial}{\partial\theta_j}\left\{(\sin\theta_j)^{N-j-1}\frac{\partial x}{\partial\theta_j}\right\},$$

and hence, setting $A = -\Lambda_N + \frac{1}{4}(N-1)(N-3)$, we have

$$(5.17)\qquad |A^{\frac{1}{2}}x|_X^2 = \sum_{j=1}^{N-1} |M_j x|_X^2 + |c_N x|_X^2 \quad (c_N^2 = \tfrac{1}{4}(N-1)(N-3)),$$

where $x(\omega)$ is a sufficiently smooth function on S^{N-1}. It follows from (5.17) that the operator M_j can be naturally extended to the operator on $D^{\frac{1}{2}}$, i.e., for $x \in D^{\frac{1}{2}}$, we define $M_j x$ by $M_j x = \operatorname{s-lim}_{n\to\infty} M_j x_n$, where $\{x_n\}$ is a sequence such that $x_n \in C^1(S^{N-1})$, $x_n \to x$ in X as $n \to \infty$ and $A^{\frac{1}{2}} x_n \to A^{\frac{1}{2}} x$ in X as $n \to \infty$. In the sequel M_j will be considered as the extended operator on $D^{\frac{1}{2}}$. Let $\lambda(y)$ be a C^2 function on $\mathbb{R}^N$ and let us set

$$(5.18)\qquad \begin{cases} \varphi = \varphi(y) = \varphi(y;\lambda) = r^{-2} \sum_{j=1}^{N-1} (M_j\lambda)^2, \\ P = P(y) = P(y;\lambda) = r^{-2}(\Lambda_N \lambda) \\ M = \sum_{j=1}^{N-1} (M_j\lambda) M_j. \end{cases}$$

<u>LEMMA 5.5</u>. Let $x \in D$ and set $\lambda(y) = \int_0^r Z(t\omega)dt$ $(r = |y|,\ \omega = y/|y|)$ with $Z(y) \in C^2(\mathbb{R}^N)$. Then we have

$$(5.19)\qquad \begin{aligned} (e^{\pm i\lambda}B(r) - B(r)e^{\pm i\lambda})x &= e^{\pm i\lambda}(-\varphi \pm 2ir^{-2}M \pm iP)x \\ &= (\varphi \pm 2ir^{-2}M \pm iP)(e^{\pm i\lambda}x), \end{aligned}$$

and

$$(5.20)\qquad \begin{aligned} (L-k^2)(e^{i\mu}x) = e^{i\mu}\{&(B(r) + 2ir^{-2}M)x + (iP + iZ' + C_1)x \\ &- (2kZ - C_0 - Z^2 - \varphi)x\}, \end{aligned}$$

where φ, P, M are given in (5.18), $Z' = \frac{\partial Z}{\partial r}$, and $\mu(y) = rk - \lambda(y)$ with $k \in \mathbb{R} - \{0\}$.

<u>PROOF</u>. (5.19) and (5.20) can be shown by easy calculation if we note

(5.21) $$\Lambda_N(e^{\pm i\lambda}x) = e^{\pm i\lambda}(\Lambda_N x \pm 2iMx \pm i(\Lambda_N\lambda)x).$$

Q.E.D.

In order to estimate the term $P(y)$, we need

LEMMA 5.6. Let $h(y) \in C^2(\mathbb{R}^N)$. Then

(5.22) $$(M_j h)(y) = \sum_{p=j}^{N} b_j(\theta)^{-1} y_{p,j} \frac{\partial h}{\partial y_p},$$

(5.23) $$(\Lambda_N h)(y) = \sum_{j=1}^{N-1} \sum_{p,q=j}^{N} b_j(\theta)^{-2} y_{p,j} y_{q,j} \frac{\partial^2 h}{\partial y_p \partial y_q} - (N-1) \sum_{p=1}^{N} y_p \frac{\partial h}{\partial y_p},$$

where $y_{p,j} = \frac{\partial y_p}{\partial \theta_j}$. Let $\lambda(y)$, $Z(y)$ be as in Lemma 5.5 and let $P(y)$ be as in (5.18). Then

(5.24) $$\begin{cases} (M_j\lambda)(y) = \int_0^r \sum_{p=j}^{N} b_j(\theta)^{-1} \left[y_{p,j} \frac{\partial Z}{\partial y_p} \right]_{y=t\omega} dt, \\ P(y) = \frac{1}{\gamma^2} \int_0^r \left\{ \sum_{j=1}^{N-1} \sum_{p,q=1}^{N} b_j(\theta)^{-2} \left[y_{p,j} y_{q,j} \frac{\partial^2 Z}{\partial y_p \partial y_q} \right]_{y=t\omega} - (N-1) \sum_{p=1}^{N} \left[y_p \frac{\partial Z}{\partial y_p} \right]_{y=t\omega} \right\} dt, \end{cases}$$

where $r = |y|$, $\omega = y/|y|$.

PROOF. Let us start with

(5.25) $$\frac{\partial h}{\partial \theta_j} = \sum_{p=j}^{N} y_{p,j} \frac{\partial h}{\partial y_p},$$

where it should be noted that $y_{p,j} = 0$ for $p < j$. (5.22) directly follows from (5.25). It follows from (5.25) that

(5.26) $$\frac{\partial^2 h}{\partial \theta_j^2} = -\sum_{p=j}^{N} y_p \frac{\partial h}{\partial y_p} + \sum_{p,q=j}^{N} y_{p,j} y_{q,j} \frac{\partial^2 h}{\partial y_p \partial y_q}.$$

Here we have used the relation

$$(5.27)\qquad \frac{\partial y_{p,j}}{\partial \theta_j} = \frac{\partial^2 y_p}{\partial \theta_j^2} = - y_p \quad (j \leqq p \leqq N, \quad 1 \leqq j \leqq N - 1).$$

Thus we obtain

$$(5.28)\qquad \begin{aligned} (\Lambda_N h)(y) &= \sum_{j=1}^{N-1} b_j(\theta)^{-2} \frac{\partial^2 h}{\partial \theta_j^2} + (N - j - 1) \frac{\cos \theta_j}{\sin \theta_j} \frac{\partial h}{\partial \theta_j} \\ &= \sum_{j=1}^{N-1} \sum_{p,q=j}^{N} b_j^{-2} y_{p,j} y_{q,j} \frac{\partial^2 h}{\partial y_p \partial y_q} - \sum_{j=1}^{N-1} \sum_{p=j}^{N} b_j^{-2} y_p \frac{\partial h}{\partial y_p} \\ &\quad + \sum_{j=1}^{N-1} \sum_{p=j}^{N} b_j^{-2} (N - j - 1) \frac{\cos \theta_j}{\sin \theta_j} y_{p,j} \frac{\partial h}{\partial y_p}, \end{aligned}$$

and, hence, we have only to show

$$(5.29)\qquad \begin{aligned} &\sum_{j=1}^{N-1} \sum_{p=j}^{N} b_j(\theta)^{-2} \left\{ (N - j - 1) \frac{\cos \theta_j}{\sin \theta_j} y_{p,j} - y_p \right\} \frac{\partial h}{\partial y_p} \\ &\qquad = - (N - 1) \sum_{p=1}^{N} y_p \frac{\partial h}{\partial y_p}. \end{aligned}$$

The order of summation in the left-hand side of (5.29) can be changed as

$$(5.30)\qquad \sum_{j=1}^{N-1} \sum_{p=j}^{N} = \sum_{p=1}^{N-1} \sum_{j=1}^{p} + \sum_{j=1}^{N}.$$

Therefore, it is sufficient to show

$$(5.31)\qquad \begin{cases} J_p = \displaystyle\sum_{j=1}^{p} b_j(\theta)^{-2} \left\{ (N-j-1) \frac{\cos \theta_j}{\sin \theta_j} y_{p,j} - y_p \right\} = - (N-1) y_p \\ \qquad\qquad\qquad\qquad\qquad\qquad (p = 1,2,\ldots,N-1) \\ J_N = \displaystyle\sum_{j=1}^{N-1} b_j(\theta)^{-2} \left\{ (N-j-1) \frac{\cos \theta_j}{\sin \theta_j} y_{p,j} - y_N \right\} = - (N-1) y_N. \end{cases}$$

To this end, let us note that

$$(5.32)\qquad \frac{\cos\theta_j}{\sin\theta_j}\, y_{p,j} = \begin{cases} \left(\dfrac{1}{\sin^2\theta_j} - 1\right) y_p & (j < p \leqq N), \\[2ex] -y_p & (j = p \leqq N-1), \end{cases}$$

whence follows that

$$\begin{aligned} J_p &= \sum_{j=1}^{p-1} b_j(\theta)^{-2}\left\{(N-j-1)\left(\frac{1}{\sin^2\theta_j} - 1\right) - 1\right\} y_p - b_p(\theta)^{-2}(N-p)y_p \\ (5.33)\qquad &= \left[\sum_{j=1}^{p-1}\left\{b_{j+1}(\theta)^{-2}(N-j-1) - b_j(\theta)^{-2}(N-j)\right\} - b_p(\theta)^{-2}(N-p)\right] y_p \\ &= -(N-1)y_p \qquad (1 \leqq p \leqq N-1). \end{aligned}$$

The relation $J_N = -(N-1)y_N$ can be proved in a quite similar way, which completes the proof of (5.23). (5.24) follows from (5.22) and (5.23). *Q.E.D.*

Let $\lambda(y)$ be as in Lemma 5.5 and let $Z(y)$ satisfy the estimate

$$(5.34)\qquad \begin{cases} D^j Z(y) = O(|y|^{-j-\varepsilon}) & (|y| \to \infty, \quad j = 0,1,2), \\[2ex] 2kZ(y) - Q_0(y) - Z(y)^2 - \varphi(y) = O(|y|^{-\tilde{\varepsilon}}) & (1 < \tilde{\varepsilon} \leqq 1+\varepsilon \\ & |y| \to \infty). \end{cases}$$

Then the estimates $(M_j\lambda)(y), (\Lambda_N\lambda)(y) = O(|y|^{1-\varepsilon})$ $(|y| \to \infty)$ are obtained from (5.24), where we should note that $|b_j(\theta)^{-1}y_{p,j}| \leqq |y|$ for $j \leqq p$. Therefore, it can be seen from (5.20) in Lemma 5.5, that $|(L-k^2)(e^{i\mu}x)|_X = O(r^{-\tilde{\varepsilon}})$ holds good under the condition (5.34). In order to obtain $\lambda(y)$, which satisfies (5.34), let us consider three sequences $\{\Phi_n(y,k)\}$, $\{\Psi_n(\omega,k)\}$, $\{\lambda_n(y,k)\}$ $(y \in \mathbb{R}^N,\ \omega \in S^{N-1},\ k \in \mathbb{R} - \{0\})$ defined by

(5.35)
$$\begin{cases}\Phi_1(y) = \frac{1}{2k}\mathcal{Q}_0(y),\\ \Psi_1(y) = 0\\ \lambda_1(y) = \frac{1}{2k}\int_0^r \mathcal{Q}_0(t\omega)dt \quad (r = |y|, \quad \omega = y/|y|),\\ \Phi_{n+1}(y) = \frac{1}{2k}\{\mathcal{Q}_0(y) + (\Phi_n(y))^2 + \varphi(y;\lambda_n(y))\} \quad (n = 1,2,\ldots,\\ \qquad\qquad (n = 1,2,\ldots,m_0-1),\\ \Psi_n(\omega) = \begin{cases} 0 & (m = 2,3,\ldots,m_1-1),\\ -\int_0^\infty\{\Phi_n(t\omega) - \Phi_{n-1}(t\omega)\}dt + \psi_{n-1}(w) & (n = m_1,m_1+1..,m_0),\end{cases}\\ \lambda_n(y) = \int_0^r \Phi_n(t\omega)dt + \xi(r)\Psi_n(\omega)\\ \qquad\qquad (r = |y|, \quad \omega = y/|y|, \quad n = 2,3,\ldots,m_0),\end{cases}$$

where $\xi(\ )$ is a real-valued C^∞-function on $\bar{I}$ such that

(5.36)
$$0 \leqq \xi(r) \leqq 1, \quad \xi'(r) \geqq 0, \quad \text{and} \quad \xi(r) = \begin{cases} 0 & (r \leqq 1),\\ 1 & (r \geqq 2),\end{cases}$$

and m_0, m_1 are as in Assumption 5.1. If $m_0 \leqq m_1$, then we set $\Psi_n(\omega) = 0$ for all $n = 1,2,\ldots,m_0-1$. In the following lemma, it can be seen that these sequences are well defined by (5.35).

<u>LEMMA 5.7</u>. (i) $\Phi_n(y,-k) = -\Phi_n(y,k)$, $\Psi_n(\omega,-k) = -\Psi_n(\omega,k)$ for any pair $(y,k) \in \mathbb{R}^N \times (\mathbb{R} - \{0\})$ and any $n = 1,2,\ldots,m_0$. Further, $\Phi_n(y,k) = 0$ for $|y| \leqq 1$ and $n = 1,2,\ldots,m_0$.

(ii) $\Phi_n(y) \in C^{m_0+1-n}(\mathbb{R}^N)$ and $\Psi_n(\omega) \in C^{m_0+1-n}(S^{n-1})$ for $n = 1,2,\ldots,m_0$.

(iii) There exists a constant $C > 0$ such that

(5.37)
$$|D^j\Phi_n(y)| \leqq C(1 + |y|)^{-j-\varepsilon}$$

and

$$(5.38)\qquad |D^j\{\Phi_n(y) - \Phi_{n-1}(y)\}| \leqq C(1 + |y|)^{-j-n\varepsilon}$$

hold for any $y \in \mathbb{R}^N$, $n = 1,2,\dots,m_0$ and $j = 0,1,2,\dots,m_0+1-n$, where D^j denotes an arbitrary derivative of the jth order.

(iv) As functions of k, Φ_n and Ψ_n are $C^{m_0+1-n}(\mathbb{R}^N)$-valued continuous functions on $\mathbb{R} - \{0\}$. The constant C in the right-hand side of (5.37) and (5.38) is bounded when k moves in a compact set in $\mathbb{R} - \{0\}$.

PROOF. (i), (ii) and (5.37), (5,38) can be shown by induction. Here it should be noted that $\mathcal{Q}_0(y) = 0$ for $|y| \leqq 1$ and

$$(5.39)\qquad (M_j h)(y) = O(|y|^{1-\tau}) \quad \text{if} \quad Dh(y) = O(|y|^{-\tau})$$
$$(\tau > 0, \quad |y| \to \infty),$$

which follows from (5.22). If $m_0 \leqq m_1$, then $\Psi_n(\omega) \equiv 0$ for all $n = 1,2,\dots,m_0-1$ and we may simply set $\lambda_n(y) = \int_0^r \Phi_n(t\omega)dt$. In the case that $m_1 < m_0$, any logarithmic term does not appear in the estimation because $(m_1-1)\varepsilon < 1$. (iv) is clear from the definitions of Φ_n and Ψ_n. Q.E.D.

DEFINITION 5.8. We set

$$(5.40)\qquad \begin{cases} Z(y) = Z(y,k) = \Phi_{m_0-2}(y) + \xi'(r)\Psi_{m_0-2}(\omega), \\ \lambda(y) = \lambda(y,k) = \lambda_{m_0-2}(y) = \int_0^r Z(t\omega)dt, \\ Y(y) = Y(y,k) = 2kZ(y) - \mathcal{Q}_0(y) - Z(y)^2 - \varphi(y;\lambda) \end{cases}$$

with $r = |y|$, $\omega = y/|y|$.

REMARK 5.9. (1) From Lemma 5.7, it can be easily seen that $Z(y) \in C^3(\mathbb{R}^N)$, $Z(y,-k) = -Z(y,k)$, $Z(y) = 0$ for $|y| \leqq 1$ and

$$(5.41)\qquad \begin{cases} |(D^jZ)(y)| \leqq C(1+|y|)^{-j-\varepsilon} & (y \in \mathbb{R}^N, \quad j = 0,1,2,3), \\ |(D^jY)(y)| \leqq C(1+|y|)^{-j-(m_0-1)\varepsilon} & (y \in \mathbb{R}^N, \quad j = 0,1). \end{cases}$$

Further, taking account of (5.5), we have

$$(5.42)\qquad |(D^jY)(y)| \leqq C(1+|y|)^{-j-2\beta-2\delta} \quad (y \in \mathbb{R}^N, \quad j = 0,1),$$

where β is given by (5.9).

(2) As a function of k, Z is a $C^3(\mathbb{R}^N)$-valued continuous function on $\mathbb{R} = \{0\}$. The constant C in (5.41) and (5.42) is bounded when k moves in a compact set in $\mathbb{R} - \{0\}$.

(3) Let us consider the case that $\frac{1}{2} < \varepsilon \leqq 1$. Then, $m_0 = 3$ and

$$(5.43)\qquad Z(y,k) = Z_1(y,k) = \frac{1}{2k}\mathcal{Q}_0(y).$$

This case was treated in Saitō [3] and Ikebe [2] (cf. (12) of Concluding Remarks).

§6. An estimate for the radiative function

Let $Z(y)$ and $\lambda(y)$ be as in §5. Now we are in a position to prove the first half of Theorem 5.4.

Let us set for a function $G(y)$ on $\mathbb{R}^N$

$$\rho_\alpha(G) = \sup_{y\in\mathbb{R}^N} (1 + |y|)^\alpha |G(y)|. \tag{6.1}$$

The following proposition is an important step to the proof of (5.10) in Theorem 5.4.

PROPOSITION 6.1. Let Assumption 5.1 be satisfied. Further assume that $\mathcal{Q}_0(y)$ has compact support in $\mathbb{R}^N$. Let K be a compact set in $\mathbb{R} - \{0\}$. Let β be as in (5.9). Then there exists $C = C(K,\mathcal{Q})$ such that

$$\|u'-iku\|_\beta + \|B^{\frac{1}{2}}u\|_\beta \leqq C\{\|f\|_{1+\beta} + \|v\|_{-\delta}\} \tag{6.2}$$

holds for any radiative function v for $\{L,k,\ell[f]\}$ with $k \in K$ and $f \in L_{2,1+\beta}(I,X)$, where $u = e^{i\lambda}v$ and $\lambda(y)$ is given in (5.40). The constant $C = C(K,\mathcal{Q})$ is bounded when $\rho_{\varepsilon_1}(\mathcal{Q}_1), \rho_{j+\varepsilon}(D^j\mathcal{Q}_0)$, $j = 0,1,2,\ldots,m_0$, are bounded. Here m_0 is given by (5.1), D^j means an arbitrary derivative of jth order and $\varepsilon,\varepsilon_1$ are given in Assumption 5.1.

In order to show this proposition, we need several lemmas.

LEMMA 6.2. Let $v \in D(I) \cap H_0^{1,B}(I,X)$ be a solution of the equation $(L-k^2)v = f$ with $k \in \mathbb{R} - \{0\}$ and $f \in L_2(I,X)_{loc}$. Set $u = e^{i\lambda}v$ with $\lambda(y)$ defined by (5.40).

(i) Then we have

$$\begin{aligned} &-(u'-iku)' - ik(u'-iku) + Bu \\ &\qquad = e^{i\lambda}f - 2iZ(u'-iku) + (Y-C_1-iZ'-iP)u - 2ir^{-2}Mu, \end{aligned} \tag{6.3}$$

where Y is as in (5.40) and P,M are given by (5.18).

(ii) Let $V(y) = \sum_{j=1}^{P} g_j G_j$ with C^1 functions g_j on $\mathbb{R}^N - \{0\}$ and operators G_j in X such that $G_j u \in C_{ac}(I,X)$ with $(G_j u)' = G_j u'$ in $L_2(I,X)_{loc}$. Then

$$\int_R^T (Vu,u'-iku)_X dr = \frac{1}{2ik} \{[(Vu,u'-iku)_X]_R^T - \int_R^T (V\{u'-iku\},u'-iku)_X dr$$

$$- \int_R^T (V'u,u'-iku)_X dr + 2i \int_R^T (ZVu,u'-iku)_X dr$$

$$(6.4) \qquad + \int_R^T (Vu,\{Y-C_1-iZ'-iP\}u + e^{i\lambda} f)_X dr$$

$$- \int_R^T (Vu,Bu)_X dr + 2i \int_R^T (Vu,r^{-2}Mu)_X dr\}$$

$$(0 < R < T)$$

with $V' = \sum_{j=1}^{P} \frac{\partial g_j}{\partial r} G_j$.

PROOF. (i) is obtained by an easy computation if we take note of Lemma 5.5 and the relation $(L-k^2)v = f$. Take the complex conjugate of both sides of (6.3), multiply it by Vu and integrate on $S^{N-1} \times (R,T)$. Then, by the use of partial integration, we arrive at (6.4). *Q.E.D.*

LEMMA 6.3. Let $x, x' \in D$ and let $S(y)$ be a C^1-function on $\mathbb{R}^N$. Then

$$(6.5) \qquad (SMx,x')_X + (Sx,Mx')_X = -r^2(SPx,x') - ((MS)x,x')_X.$$

PROOF . By partial integration, we have

$$(6.6) \qquad (Sx,(M_j\lambda)M_j x')_X = \int_{S^{N-1}} S(r\omega)x(\omega)(M_j\lambda)(r\omega)b_j(\theta)^{-1} \frac{\partial \overline{x'}}{\partial \theta_j} d\omega$$

$$(d\omega = (\sin\theta_1)^{N-2} \ldots (\sin\theta_j)^{N-j-1} \ldots \sin\theta_{N-2})$$

$$= - \int_{S^{N-1}} \frac{\partial}{\partial \theta_j}\{S(r\omega)x(\omega)(M_j\lambda)(r\omega)(\sin\theta_j)^{N-j-1}\}b_j(\theta)^{-1}\overline{x}'(\omega)\cdot$$

$$\cdot \frac{d\omega}{(\sin\theta_j)^{N-j-1}}$$

$$= - \int_{S^{N-1}} \{(M_j\lambda)M_j(Sx) + Sxb_j(\theta)^{-2}(\sin\theta_j)^{-N-j-1}\ \cdot$$

$$\cdot \frac{\partial}{\partial\theta_j}((\sin\theta_j)^{N-j-1}\frac{\partial}{\partial\theta_j})\overline{x}'\}d\omega$$

whence we obtain

$$(6.7)\qquad \begin{aligned}(S\ ,Mx')_X &= -(M(Sx) + Sx(\Lambda_N\lambda),x')_X \\ &= -(SMx,x')_X - ((MS)x,x')_X = r^2(SPx,x')_X.\end{aligned}$$

This is what we wanted to show. *Q.E.D.*

LEMMA 6.4. Let $\mathcal{Q}_0(y)$ be as in Proposition 6.1. Let v be the radiative function for $\{L,k,\ell[f]\}$ with $k \in \mathbb{R} - \{0\}$ and $f \in L_{2,1+\beta}(I,X)$, where β is as in (5.9), i.e., $\beta = \delta - \varepsilon$ $(0 < \varepsilon \leq \frac{1}{2})$, $= 0$ $(\frac{1}{2} < \varepsilon \leq 1)$. Then we have $v'-ikv, B^{\frac{1}{2}}v \in L_{2,\beta}(I,X)$

PROOF. Let $k_n = k + \frac{i}{n}$ $(n = 1,2,3,\ldots)$ and let v_n be the radiative function for $\{L,k,\ell[f]\}$. From Lemma 2.7, it follows that $v_n'-ikv_n, B^{\frac{1}{2}}v_n \in L_{2,1+\beta}(I,X)$. Multiply both sides of

$$(6.8)\qquad \begin{aligned}-((v_n'-ikv_n)',v_n'-ik_nv_n)_X - ik_n|v_n'-ik_nv_n|_X^2 + (Bv_n,v_n'-ik_nv_n)_X \\ + (Cv_n,v_n'-ik_nv_n)_X = (f,v_n'-ikv_n)_X\end{aligned}$$

by $\xi(r)(1 + r)^{2\beta+1}$, $\xi(r)$ being given by (5.36), take the real part and integrate from 0 to T. Then, proceeding as in the proof of Lemma 2.5 and using (5.5), we arrive at

$$(\beta + \tfrac{1}{2}) \int_0^T \xi r^{2\beta} |v_n' - ik_n v_n|_X^2 dr + (\tfrac{1}{2} - \beta) \int_0^T \xi r^{2\beta} |B^{\frac{1}{2}} v_n|_X^2 dr$$

$$\leqq \tfrac{1}{2}(1+T)^{2\beta+1} |v_n'(T) - ik_n v_n(T)|_X^2$$

$$(6.9) \qquad + \tfrac{1}{2} \int_0^T \xi'(1+r)^{2\beta+1} \{|B^{\frac{1}{2}} v_n|_X^2 - |v_n' - ik_n v_n|_X^2\} dr$$

$$+ C \int_0^T \xi(1+r)^{-\delta+\beta} |v_n' - ik_n v_n|_X |v_n|_X dr$$

$$+ \int_0^T \xi(1+r)^{2\beta+1} |f|_X |v_n' - ik_n v_n|_X dr,$$

whence, by the use of Proposition 1.2 (the interior estimate) and Theorem 2.3 (the limiting absorption principle), it follows that

$$(6.10) \qquad \|v_n' - ik_n v_n\|_\beta + \|B^{\frac{1}{2}} v_n\|_\beta \leqq C\|f\|_{1+\beta} \qquad (n = 1,2,\ldots)$$

with a constant C which is independent of $n = 1,2,\ldots$. Let $n \to \infty$ in the estimate

$$(6.11) \qquad \|v_n' - ikv_n\|_{\beta,(0,R)} + \|B^{\frac{1}{2}} v_n\|_{\beta,(0,R)} \leqq C\|f\|_{1+\beta}.$$

Then it follows from Theorem 2.3 that

$$(6.12) \qquad \|v' - ikv\|_{\beta,(0,R)} + \|B^{\frac{1}{2}} v\|_{\beta,(0,R)} \leqq C\|f\|_{1+\beta}$$

holds for any arbitrary $R > 0$, which implies that $v' - ikv, B^{\frac{1}{2}} v \in L_{2,\beta}(I,X)$. *Q.E.D.*

<u>LEMMA 6.5</u>. Let $Q(y)$ be as in Proposition 6.1 and let K be a compact set in $\mathbb{R} - \{0\}$. Let v be the radiative function for $\{L,k,\ell[f]\}$ with $k \in K$ and $f \in L_{2,1+\beta}(I,X)$, β being as in (5.9). Let $\alpha(r) = \alpha_R(r) = \xi(r-R+1)$, where $\xi(r)$ is given by (5.36). Then we have for $T \geqq R+1 > R \geqq 1$

$$\int_R^T \alpha r^{2\beta}(|u'-iku|_X^2+|B^{\frac{1}{2}}u|_X^2)dr \leqq \eta(T) + C_R \int_R^{R+1} (|u|_X^2+|u'-iku|_X^2+|B^{\frac{1}{2}}u|_X^2)dr$$

$$(6.13) \qquad + C_1\left\{\int_R^T \alpha r^{2\beta-\varepsilon}(|u'-iku|_X^2 + |B^{\frac{1}{2}}u|_X^2)dr + \|f\|_{1+\beta}^2 + \|v\|_{-\delta}^2\right\}$$

$$+ C_2\left[\sum_{n=0}^{m_1-1} k^{-n-1}\mathrm{Re}\int_R^T \alpha r^{2\beta-1}(Z^n Mu,Bu)_X dr\right],$$

where $u = e^{i\lambda}v$, $\eta(T)$ is a function of T satisfying $\lim_{T\to\infty}\eta(T) = 0$, m_1 is as in Assumption 5.1, C_R is a constant depending only on R and K, and $C_p = C_p(K,\mathcal{Q})$, $p = 1,2$, are bounded when $\rho_{\varepsilon_1}(\mathcal{Q}_1),\rho_{j+\varepsilon}(D^j\mathcal{Q}_0)$, $j = 0,1,2,\ldots,m_0$, are bounded.

<u>PROOF</u>. Multiply both sides of (6.3) by $\alpha r^{2\beta+1}(\overline{u'-iku})$, integrate over the region $S^{N-1}\times(R,T)$ and take the real part. Then we have

$$(6.14) \qquad \begin{aligned} K &= \mathrm{Re}\int_R^T \alpha r^{2\beta+1}\{-((u'-iku)',u'-iku)_X + (Bu,u'-iku)_X\}dr \\ &= \mathrm{Re}\int_R^T \alpha r^{2\beta+1}\{(e^{i\lambda}f,u'-iku)_X + ((Y - C_1)u,u'-iku)_X\}dr \\ &\quad + \mathrm{Im}\int_R^T \alpha r^{2\beta+1}((Z' + P)u,u'-iku)_X dr \\ &\quad + 2\mathrm{Im}\int_R^T \alpha r^{2\beta-1}(Mu,u'-iku)_X dr = K_1 + K_2 + K_3. \end{aligned}$$

The left-hand side K of (6.14) is estimated from below as follows:

$$(6.15) \qquad \begin{aligned} K &\geqq (\beta + \tfrac{1}{2})\int_R^T \alpha r^{2\beta}|u'-iku|_X^2 dr + (\tfrac{1}{2} - \beta)\int_R^T \alpha r^{2\beta}|B^{\frac{1}{2}}u|_X^2 dr \\ &\quad + \frac{1}{2}\int_R^{R+1} \alpha' r^{2\beta+1}\{|u'-iku|_X^2-|B^{\frac{1}{2}}u|_X^2\}dr - \tfrac{1}{2}T^{2\beta+1}|u'(T)-iku(T)|_X^2. \end{aligned}$$

K_1 can be estimated as

$$(6.16) \qquad K_1 \leqq C\{\|f\|_{1+\beta}\left[\int_R^T \alpha r^{2\beta}|u'-iku|_X^2 dr\right]^{\frac{1}{2}} + \|f\|_{1+\beta}^2 + (R+1)^{1-2\delta}\int_R^{R+1}\alpha'|u|_X^2 dr + \|v\|_{-\delta}^2$$

$$+ T^{1-2\delta}|u(T)|_X^2\},$$

where we integrated by parts and used (5.42) in estimating the term $\mathrm{Re}\int_R^T \alpha r^{2\beta+1}(Yu,u'-iku)_X dr$. Here and in the sequel, the constant depending only on K and $\mathcal{Q}(y)$ will be denoted the same symbol C.

It follows from (6.14) - (6.16) that

$$c_1 \int_R^T \alpha r^{2\beta}(|u'-iku|_X^2 + |B^{\frac{1}{2}}u|_X^2)dr$$

$$\leq C\{T^{2\delta+1}|u'-iku|_X^2 + T^{1-2\delta}|u(T)|_X^2 + (1+R)^{2\beta+1}\int_R^{R+1}\alpha'|B^{\frac{1}{2}}u|_X^2 dr \tag{6.17}$$

$$+ \|v\|_{-\delta}^2 + \int_R^{R+1}\alpha'|u|_X^2 dr + \|f\|_{1+\beta}^2\} + K_2 + K_3 \quad (c_1 = \tfrac{1}{2}(\tfrac{1}{2}-\beta)).$$

In order to estimate K_2 and K_3 in (6.17), we have to make use of Lemma 6.3 (ii). Set $V = \alpha r^{2\beta+1}(Z' + P)$ in (6.4). Then

$$K_2 \leq -\frac{1}{2}\mathrm{Re}\{T^{2\beta+1}((Z'(T)+P(T))u(T),u'(T)-iku(T))_X\} + F(T) + G(R)$$

$$+ \frac{1}{k}\mathrm{Im}\int_R^T \alpha r^{2\beta+1}(Z(Z'+P)u,u'-iku)_X dr \tag{6.18}$$

$$+ \frac{1}{k}\mathrm{Im}\int_R^T \alpha r^{2\beta-1}((Z'+P)u,Mu)_X dr,$$

where $F(T)$ and $G(R)$ are the terms of the form

$$\begin{cases} F(T) = C\{\|f\|_{1+\beta}^2 + \int_R^T \alpha r^{2\beta-\varepsilon}(|u'-iku|_X^2 + |B^{\frac{1}{2}}u|_X^2)dr + \|v\|_{-\delta}^2\}, \\ G(R) = C_R \int_R^{R+1}(|u|_X^2 + |u'-iku|_X^2 + |B^{\frac{1}{2}}u|_X^2)dr\}, \end{cases} \tag{6.19}$$

C_R being a constant depending only on R and K. Here, (5.17) is necessary to estimate the term $((Z' + P)u,Bu)_X$. Let us next set $V = \alpha r^{2\beta-1}M$ in (6.4). Then it follows that

$$K_3 \leq -\frac{1}{k}\,\mathrm{Re}\{T^{2\beta-1}(Mu(T),u'(T)-iku(T))_X\} + F(T) + G(R)$$

(6.20)
$$+\frac{2}{k}\,\mathrm{Im}\int_R^T \alpha r^{2\beta-1}(ZMu,u'-iku)_X dr$$
$$+\frac{1}{k}\,\mathrm{Im}\int_R^T \alpha r^{2\beta-1}(Mu,(Z'+P)u)_X dr$$
$$+\frac{1}{k}\,\mathrm{Re}\int_R^T \alpha r^{2\beta-1}(Mu,Bu)_X dr$$

Here we used the relation $\mathrm{Re}(M(u'-iku),u'-iku)_X = -(r^2/2)(P(u'-iku),u'-iku)_X$ which follows from (6.5) with $x = x' = u'-iku$ and $S(y) = 1$. Lemma 6.3 was also used in order to estimate the term $\mathrm{Re}(Mu,Yu)_X$. Thus, we obtain from (6.17), (6.18) and (6.20)

(6.21)
$$\int_R^T \alpha r^{2\beta}(|u'-iku|_X^2 + |B^{\frac{1}{2}}u|_X^2)dr \leq \eta_1(T) + F(T) + G(R)$$
$$+ C\{\frac{1}{k}\,\mathrm{Im}\int_R^T \alpha r^{2\beta+1}(Z(Z'+P)u,u'-iku)_X dr$$
$$+\frac{2}{k}\,\mathrm{Im}\int_R^T \alpha r^{2\beta-1}(ZMu,u'-iku)_X dr + \frac{1}{k}\,\mathrm{Re}\int_R^T \alpha r^{2\beta-1}(Mu,Bu)_X dr\}$$
$$= \eta_1(T) + F(T) + G(R) + C\{J_2 + J_3 + J_4\},$$

where we have used the fact that $\mathrm{Im}\{((Z'+P)u,Mu)_X + (Mu,(Z'+P)u)_X\} = 0$, and

(6.22)
$$\eta_1(T) = C\{T^{2\beta+1}|u'(T)-iku(T)|_X^2 + T^{1-2\delta}|u(T)|_X^2 + T^{2\beta+1}|B^{\frac{1}{2}}u(T)|_X^2\}.$$

Since $v'-iku, B^{\frac{1}{2}}v \in L_{2,\beta}(I,X)$ by Lemma 6.4 and the support of $Z(y)$ is compact in $\mathbb{R}^N$ by the compactness of $\mathcal{Q}_0(y)$, it can be easily seen that $u'-iku, B^{\frac{1}{2}}u \in L_{2,\beta}(I,X)$, which implies that $\underline{\lim}_{T\to\infty}\,\eta(T) = 0$. J_2 and J_3 in (6.21) can be estimated in quite the same way as in the estimation of K_2 and K_3 respectively. By repeating these estimations, we arrive at

(6.23)
$$\int_R^T \alpha r^{2\beta}(|u'-iku|_X^2 + |B^{\frac{1}{2}}u|_X^2)dr$$

$$\leq \eta(T) + F(T) + G(R)$$

$$+ C\left\{\sum_{n=0}^{m_1-1} k^{-n-1} \operatorname{Re} \int_R^T \alpha r^{2\beta-1}(Z^n Mu, Bu)_X dr\right.$$

$$+ k^{-m_1} \operatorname{Im} \int_R^T \alpha r^{2\beta+1}(Z^{m_1}(Z'+P)u, u'-iku)_X dr$$

$$\left. + 2k^{-m_1} \operatorname{Im} \int_R^T \alpha r^{2\beta-1}(Z^{m_1} Mu, u'-iku)_X dr\right\},$$

where $\eta(T)$ is the term satisfying $\varlimsup_{T\to\infty} \eta(T) = 0$. By noting that $Z(y)^{m_1} = O(|y|^{-1})$, (6.13) is obtained from (6.23). *Q.E.D.*

We have only to estimate the terms $\operatorname{Re}(Z^n Mu, Bu)_X$ in order to show Proposition 6.1 completely.

LEMMA 6.6. Let $S(y)$ be a real-valued C^1 function such that

$$\text{(6.24)} \qquad \begin{cases} |S(y)| \leq c & (y \in \mathbb{R}^N), \\ |DS(y)| \leq cr^{-1} & (|y| \geq 1) \end{cases}$$

with a constant $c > 0$, where $D = \dfrac{\partial}{\partial y_j}$ $(j = 1,2,\ldots,N)$. Then the estimate

$$\text{(6.25)} \qquad |\operatorname{Re}(S\,Mx, Ax)_X| \leq Cr^{1-\varepsilon}(|A^{\frac{1}{2}}x|_X^2 + |x|_X^2) \quad (r \geq 1,\ \ k \in K,\ \ x \in D)$$

holds with $C = C(c, K, \mathcal{Q}_0)$ which is bounded when c and $\rho_{j+\varepsilon}(\mathcal{Q}_0)$, $j = 0,1,2,\ldots,m_0$, are bounded. K is a compact set in $\mathbb{R} - \{0\}$ and $A = -\Lambda_N + \frac{1}{4}(N-1)(N-3)$.

PROOF. We shall divide the proof into several steps.

(I) By the use of (5.17) $J = \operatorname{Re}(SMx, Ax)_X$ can be rewritten as

$$\text{(6.26)} \qquad \begin{aligned} J = \operatorname{Re}(SMx, Ax)_X &= \operatorname{Re}\left\{\sum_{n=1}^{N-1}(M_n(SMx), M_n x)_X\right\} \\ &+ c_N^2 \operatorname{Re}(SMx, x)_X = J_1 + J_2. \end{aligned}$$

Throughout this proof we shall call a term K an O.K. term when K is dominated by $Cr^{1-\varepsilon}(|A^{\frac{1}{2}}x|_X^2 + |x|_X^2)$ for $r \geq 1$ with $C = C(c,K,Q)$. Since $|Mx|_X \leq Cr^{1-\varepsilon}|A^{\frac{1}{2}}x|_X$, we can easily show that J_2 is an O.K. term. Thus, we have only to consider the term J_1.

(II) Let us calculate J_1.

$$J_1 = \mathrm{Re}\left\{\sum_{n=1}^{N-1}((M_n S)Mx, M_n x)_X\right\}$$

(6.27)
$$+ \mathrm{Re}\left\{\sum_{n=1}^{N-1}(S\sum_{j=1}^{N-1}(M_n M_j \lambda)(M_j x), M_n x)_X\right\}$$

$$+ \mathrm{Re}\left\{\sum_{n=1}^{N-1}(S\sum_{j=1}^{N-1}(M_j \lambda)(M_n M_j x), M_n x)_X\right\} = J_{11} + J_{12} + J_{13}.$$

By noting that $M_n S(y)$ is bounded on $\{y \in \mathbb{R}^N / |y| \geq 1\}$, J_{11} is seen to be an O.K. term. Before calculating J_{13}, we mention

(6.28)
$$M_n M_j - M_j M_n = \begin{cases} b_j^{-1} \dfrac{\cos\theta_j}{\sin\theta_j} M_n & (n > j), \\ 0 & (n = j), \\ -b_n^{-1} \dfrac{\cos\theta_n}{\sin\theta_n} M_j & (n < j), \end{cases}$$

which is obtained directly from the definition of M_j (see (5.15)). Using (6.28), we have

$$J_{13} = \mathrm{Re}\left[\sum_{n=1}^{N-1}\left\{(S\sum_{j=1}^{n-1}(M_j\lambda)b_j^{-1}\frac{\cos\theta_j}{\sin\theta_j}(M_n x), M_n x)_X\right.\right.$$

(6.29)
$$\left.\left. - (S\sum_{j=n+1}^{N-1}(M_j\lambda)b_n^{-1}\frac{\cos\theta_n}{\sin\theta_n}(M_j x), M_n x)_X\right\}\right]$$

$$+ \mathrm{Re}\left\{\sum_{n=1}^{N-1}(SM(M_n x), M_n x)_X\right\} = J_{131} + J_{132}.$$

Here, J_{132} is an O.K. term, because, making use of (6.5) with $x = x' = M_n x$, we have

(6.30) $$J_{132} = -\frac{1}{2}\sum_{n=1}^{N-1}\{r^2(SP(M_n x),(M_n x))_x + ((MS)M_n x, M_n x)_x\}.$$

Therefore, let us consider $J_{12} + J_{131}$.

(III) Set

(6.31) $$\begin{cases} Z_p(r\omega) = \int_0^r \left[\frac{\partial Z}{\partial y_p}\right]_{y=t\omega} t\,dt & (p = 1,2,\ldots,N), \\ Z_{jn}(r\omega) = (b_j b_n)^{-1} \sum_{\substack{p=j\\q=n}}^{N} \int_0^r \left[\frac{\partial^2 Z}{\partial y_p \partial y_q} y_{p,j} y_{q,n}\right]_{y=t\omega} dt & (j,n = 1,2,\ldots,N-1), \end{cases}$$

where $y_{p,j} = \frac{\partial y_p}{\partial \theta_j}$. Then, setting $\cos\theta_N \equiv 1$ and starting with (5.24), we obtain

(6.32) $$M_j\lambda = b_j^{-1}\{-b_{j+1}Z_j + \sum_{p=j+1}^{N} b_p Z_p \frac{\cos\theta_j}{\sin\theta_j}\cos\theta_p\},$$

and

(6.33) $$M_n M_j \lambda = \begin{cases} Z_{jn} & (j > n), \\ Z_{nn} - b_n^{-2} \sum_{p=n}^{N} Z_p b_p \cos\theta_p & (j = n), \\ Z_{jn} + b_j^{-1} \frac{\cos\theta_j}{\sin\theta_j}(M_n\lambda) & (j < n). \end{cases}$$

Then J_{12} is rewritten as

(6.34) $$\begin{aligned} J_{12} = \mathrm{Re}&\left\{\sum_{n=1}^{N-1}(S\sum_{j=1}^{n-1} b_j^{-1}\frac{\cos\theta_j}{\sin\theta_j}(M_n\lambda)(M_j x), M_n x)_x\right. \\ &\left.- \sum_{n=1}^{N-1}(S b_n^{-2}\sum_{p=n}^{N} Z_p b_p \cos\theta_p (M_n x), M_n x)_x\right\} \\ &+ \mathrm{Re}\left\{\sum_{n=1}^{N-1}(S\sum_{j=1}^{n-1} Z_{jn}(M_j x), M_n x)_x\right\} = J_{121} + J_{122}. \end{aligned}$$

Here, J_{122} is an O.K. term because $Z_{jn}(y) = O(|y|^{1-\varepsilon})$ by the first estimate of (5.34). Hence, in place of $J_{12} + J_{131}$, it is sufficient to consider

$$J' = J_{121} + J_{131}$$

(6.35)

$$= \mathrm{Re}\left\{\sum_{n=1}^{N-1}(SF_n(M_n x), M_n x)_x\right\} + \mathrm{Re}\left\{\sum_{n=1}^{N-1}(SG_n, M_n x)_x\right\} = J_1' + J_2'$$

with

(6.36)

$$\begin{cases} F_n = \sum_{j=1}^{n-1}(M_j\lambda)b_j^{-1}\dfrac{\cos\theta_j}{\sin\theta_j} - \sum_{p=n}^{N} Z_p b_p b_n^{-2}\cos\theta_p = F_{n1} - F_{n2}, \\[2ex] G_n = \sum_{j=1}^{n-1} b_j^{-1}\dfrac{\cos\theta_j}{\sin\theta_j}(M_n\lambda)(M_j x) - \sum_{j=n+1}^{N-1} b_n^{-1}\dfrac{\cos\theta_n}{\sin\theta_n}(M_j\lambda)(M_j x). \end{cases}$$

(IV) Now let us calculate J'. Using (6.32) and interchanging the order of summation, we arrive at

(6.37)
$$F_{n1} = -\sum_{p=1}^{N} Z_p b_p \cos\theta_p + \sum_{p=n}^{N} b_n^{-2} b_p Z_p \cos\theta_p,$$

and hence

(6.38)
$$F_n = -\sum_{p=1}^{N} Z_p b_p \cos\theta_p,$$

which implies that J_1' is an O.K. term. As for J_2', we can interchange the order of summation to obtain

(6.39)

$$J_2' = \mathrm{Re}\left[\sum_{n=1}^{N-1}\sum_{j=1}^{N-1}\left\{(Sb_j^{-1}\frac{\cos\theta_j}{\sin\theta_j}(M_n\lambda)(M_j x), M_n x)_x\right.\right.$$

$$\left.\left. - (Sb_j^{-1}\frac{\cos\theta_j}{\sin\theta_j}(M_n\lambda)(M_n x), M_j x)_x\right\}\right]$$

$$= 0.$$

Thus, we have shown that each term of $J = \mathrm{Re}(SMx, Ax)_X$ is an O.K. term, which completes the proof. *Q.E.D.*

Proof of Proposition 6.1. By Lemma 6.6, the last term of the right-hand side of (6.13) is dominated by the term of the form $F(T)$, where $F(T)$ is given in (6.19). Therefore, by letting $T \to \infty$ along a suitable sequence $\{T_n\}$ in (6.13), we obtain

$$\int_R^\infty \alpha^{2\beta}(|u'-iku|_X^2 + |B^{\frac{1}{2}}u|_X^2)dr \leqq G(R) + C\left\{\int_R^\infty \alpha r^{2\beta-\varepsilon}(|u'-iku|_X^2 + |B^{\frac{1}{2}}u|_X^2)dr + \|f\|_{1+\beta}^2 + \|v\|_{-\delta}^2\right\}, \tag{6.40}$$

where $C = C(K,\mathcal{Q})$ is independent of $R > 0$. Take $R = R_0$ sufficiently large in (6.40). Then it follows that

$$\int_{R_0+1}^\infty r^{2\beta}(|u'-iku|_X^2 + |B^{\frac{1}{2}}u|_X^2)dr \leqq C\{\|f\|_{1+\beta}^2 + \|v\|_{-\delta}^2\} + G(R_0) \quad (C = C(K,\mathcal{Q})). \tag{6.41}$$

Since $G(R_0)$ is dominated by the term of the form $C\{\|f\|_{1+\beta}^2 + \|v\|_{-\delta}^2\}$ by using Proposition 1.2, (6.2) easily follows from (6.41). Re-examining the proof of Lemmas 6.2 - 6.6, we can easily see that $C = C(K,\mathcal{Q})$ remains bounded when $\rho_{\varepsilon_1}(\mathcal{Q}_1), \rho_{j+\varepsilon}(D^j\mathcal{Q}_0)$, $j = 0,1,2,\dots,m_0$ are bounded. *Q.E.D.*

Now that Proposition 6.1 has been shown, we can prove (5.10) in Theorem 5.4.

<u>PROOF of (5.10) in THEOREM 5.4.</u> Let v be the radiative function for $\{L,k,\ell\}$ with $k \in K$, the compact set of $\mathbb{R} - \{0\}$, and $\ell \in F_{1+\beta}$, where $\delta - 1 \leqq \beta \leqq 1 - \delta$. Let $k_0 \in \mathbb{C}^+$ such that $\mathrm{Im}\, k_0 > 0$. Then v can be decomposed as $v = v_0 + w$, where v_0 is the radiative function for $\{L,k_0,\ell\}$ and w is the radiative function for $\{L,k,\ell[f]\}$, $f = (k^2-k_0^2)v_0$.

It follows from Lemma 2.7 that

(6.42) $$\|v_0'\|_{1+\beta} + \|B^{\frac{1}{2}}v_0\|_{1+\beta} + \|v_0\|_{1+\beta} \leqq C_0|||\ell|||_{1+\beta} \quad (C_0 = C_0(k_0,\beta)).$$

Set $u_0 = e^{i\lambda}v_0$. Noting that

(6.43) $$u_0' - iku_0 = e^{i\lambda}(v_0' - ikv_0) + iZe^{i\lambda}v_0,$$

and

(6.44) $$\begin{aligned} 0 \leqq (B(r)u_0,u_0)_X &= (B(r)v_0,v_0)_X - (\{iP + 2ir^{-2}M-\varphi\}v_0,v_0)_X, \\ &\leqq |B^{\frac{1}{2}}v_0|_X^2 + C\{|v_0|_X^2 + |B^{\frac{1}{2}}v_0|^2\} \end{aligned}$$

with $C = C(K)$, which follows from (5.19) in Lemma 5.5, we obtain from (6.42)

(6.45) $$\|u_0'-iku_0\|_\beta + \|B^{\frac{1}{2}}u_0\|_\beta \leqq C|||\ell|||_{\beta+1} \quad (C = C(k_0,K,\beta)).$$

Therefore, it suffices to show the estimate (5.10) with $u = e^{i\lambda}w$. To this end, we shall approximate $\mathcal{Q}_0(y)$ by a sequence $\{\mathcal{Q}_{0n}(y)\}$, where we set

(6.46) $$\mathcal{Q}_{0n}(y) = \rho\left(\frac{|y|}{n}\right)\mathcal{Q}_0(y) \qquad (n = 1,2,\ldots),$$

and $\rho(r)$ is a real-valued, smooth function on $\overline{I}$ such that $\rho(r) = 1$ $(r \leqq 1)$, $= 0$ $(r \geqq 2)$. Then it can be easily seen that $\rho_{j+\varepsilon}(D^j\mathcal{Q}_{0n})$ is bounded uniformly for $n = 1,2,\ldots$ with $j = 0,1,2,\ldots,m$. Let us set

(6.47) $$L_n = -\frac{d^2}{dr^2} + B(r) + C_{0n}(r) + C_1(r) \qquad (C_{0n}(r) = \mathcal{Q}_{0n}(r\omega)x),$$

and let us denote by w_n the radiative function for $\{L_n,k,\ell[f]\}$ with $f = (k^2 - k_0^2)v_0$ $(n = 1,2,\ldots)$. For each L_n, the function $Z^{(n)}(y)$

can be constructed according to Definition 5.8 with $\mathcal{Q}_0(y)$ replaced by $\mathcal{Q}_{0n}(y)$, and we set

$$(6.48) \qquad u_n = e^{i\lambda^{(n)}} w_n \qquad (\lambda^{(n)}(r\omega) = \int_0^r Z^{(n)}(t\omega)dt).$$

Now, Proposition 6.1 can be applied to show

$$(6.49) \qquad \|u_n'-iku_n\|_\beta + \|B^{\frac{1}{2}}u_n\|_\beta \leqq C\{\|f\|_{1+\beta}+\|w_n\|_{-\delta}\} \qquad (n = 1,2,\ldots,),$$

with $C = C(K)$. Since $D^j\mathcal{Q}_{0n}(y)$ converges to $D^j\mathcal{Q}_0(y)$ as $n \to \infty$ uniformly on $\mathbb{R}^N$ for each $j = 0,1,\ldots,m_0$, it follows that $\lambda^{(n)}(y) \to \lambda(y)$ as $n \to \infty$ uniformly on every compact set in $\mathbb{R}^N$. Therefore, by the use of Theorem 4.5, we obtain $u_n \to u$ in $H^{1,\beta}_{0-\delta}(I,X)$ as $n \to \infty$. Let $n \to \infty$ in the relation

$$(6.50) \qquad \|u_n'-iku_n\|_{\beta,(0,R)} + \|B^{\frac{1}{2}}u_n\|_{\beta,(0,R)} \leqq C\{\|f\|_{1+\beta} + \|w_n\|_{-\delta}\}$$

with $R > 0$, which is a direct consequence of (6.49). Then we have

$$(6.51) \qquad \|u'-iku\|_{\beta,(0,R)} + \|B^{\frac{1}{2}}u\|_{\beta,(0,R)} \leqq C\{\|f\|_{1+\beta} + \|v\|_{-\delta}\}.$$

Since $R > 0$ is arbitrary, we have obtained

$$(6.52) \qquad \|u'-iku\|_\beta + \|B^{\frac{1}{2}}u\|_\beta \leqq C\{\|f\|_{1+\beta} + \|v\|_{-\delta}\}.$$

(5.10) follows from (6.42), (6.45), (6.52) and (2.7). Q.E.D.

§7. Proof of the main theorem

This section is devoted to showing (5.11) in Theorem 5.4 and Theorem 5.3 by the use of (5.10) in Theorem 5.4 which has been proved in the preceding section.

<u>PROOF of (5.11) in THEOREM 5.4.</u> Let us first consider the case that v is the radiative function for $\{L,k,\ell[f]\}$ with $k \in K$ and $f \in L_{2,1+\beta}(I,X)$. Using (i) of Lemma 6.2, we have for $u=e^{i\lambda}v$

$$(7.1)\quad \begin{cases} \frac{d}{dr}\{e^{2irk}(u'(r)-iku(r),u(r))_x\}=e^{2ikr}g(r), \\ g(r)=|u'-iku|_x^2-|B^{\frac{1}{2}}u|_x^2-(e^{i\lambda}f,u)_x+2i(Z(u'-iku),u)_x \\ \quad -((Y-C,-iZ'-ip)u,u)_x+2ir^{-2}(Mu,u)_x. \end{cases}$$

It follows from (5.10) that $g(r)$ is integrable over I with the estimate

$$\int_I|g(r)|_x dr \leq \|u'-iku\|_0^2+\|B^{\frac{1}{2}}u\|_0^2+\|f\|_\delta\|v\|_{-\delta}$$

$$(7.2)\quad +C\{\|u'-iku\|_\beta\|v\|_{-\delta}+\|v\|_{-\delta}^2+\|B^{\frac{1}{2}}u\|_\beta\|v\|_{-\delta}\}$$

$$\leq C'\|f\|_{1+\beta}^2 \qquad (\beta=\delta-\varepsilon(0<\varepsilon\leq\frac{1}{2}),=0(\frac{1}{2}<\varepsilon\leq 1)),$$

Where $C=C(K)$, $C'=C(K)$ and we have made use of (2.7), too. Integrate the first relation of (7.1) from r to R ($0 \leq r < R < \infty$), multiply the result by e^{-2irk} and take the imaginary part. Then we have

$$k|v(r)|_x^2=k|u(r)|_x^2$$

$$(7.3)\quad =I_m(u'(r),u(r))_x-I_m\{e^{2ik(R-r)}(u'(R)-iku(R),u(R))_x\}$$

$$+Im\int_r^R e^{2ik(t-r)}g(t)dt.$$

By letting $R \to \infty$ along a sequence $\{R_n\}$ such that $(u'(R_n)-iku(R_n),u(R_n))_x \to 0$ ($n \to \infty$) (7.3) gives

$$(7.4)\quad |v(r)|_x^2 \leq \frac{1}{|k|}\left\{|I_m(u'(r),u(r))_x|+\int_r^\infty|g(t)|dt\right\}.$$

Let us estimate $I_m(u'(r),u(r))_x$. Since

$$(7.5)\quad \begin{aligned}&\frac{d}{dr}\{I_m(u'(r),u(r))_x\}=I_m\{\frac{d}{dr}(u'(r),u(r))_x\}\\ &=I_m\left[e^{-2ikr}\frac{d}{dr}\{e^{2irk}(u'(r)-iku(r),u(r))_x\}\right]=I_mg(r),\end{aligned}$$

and

(7.6) $\quad I_m(u'(r),u(r))_X=(Z(r\cdot)v(r),v(r))_X+I_m(v'(r),v(r))_X\to 0$

$(r\to 0)$

by (3.36), it follows that

(7.7) $\quad |I_m(u'(r),u(r))_X|\leq \int_0^r |g(t)|dt.$

Thus we obtain from (7.4), (7.7) and (7.2)

(7.8) $\quad |v(r)|_X\leq C\|f\|_{1+\beta} \qquad (r\in I)$

with C=C(K). Let us next consider the general case. v is now assumed to be the radiative function for $\{L,k,\ell\}$ with $k\in K$ and $\ell\in F_{1+\beta}(I,X)$. Then, according to Lemma 2.8, we decompose v as $v=v_0+w$, where v_0 is the radiative function for $\{L,k_0,\ell\}$ with $k_0\in\mathbb{C}^+$, $I_m k_0>0$ and w is the radiative function for $\{L,k,\ell[(k^2-k_0^2)v_0]\}$. It follows from Lemma 2.7 and (1.8) in Proposition 1.1 that

(7.9) $\quad \begin{cases} |v_0(r)|\leq\sqrt{2}\|v_0\|_\beta\leq C|||\ell|||_{1+\beta} & (r\in I), \\ \|v_0\|_{1+\beta}\leq C|||\ell|||_{1+\beta} \end{cases}$

with $C=C(k_0)$. On the other hand we obtain from (7.8)

(7.10) $\quad |w(r)|_X\leq C\|v_0\|_{1+\beta} \qquad (r\in I)$

with $C=C(K,k_0)$. (5.11) directly follows from (7.9) and (7.10).

Q.E.D.

The proof of Theorem 5.3 will be divided into the following four steps. Lemmas 7.1, 7.2, 7.4 and Corollary 7.3 will be concerned with the asymptotic behavior of the radiative function v for $\{L,k,\ell[f]\}$ with $k\in\mathbb{R}-\{0\}$ and $f\in L_{2,1+\beta}(I,X)$.

LEMMA 7.1. Let v be the radiative function for $\{L,k,\ell[f]\}$ with $k\in\mathbb{R}-\{0\}$ and $f\in L_{2,1+\beta}(I,X)$. Then $|v(r)|_X$ tends to a limit as $r\to\infty$.

PROOF. Let $R_0>0$ be fixed. (7.3) are combined with (7.5) to give

$$(7.11)\quad |v(r)|_X^2=\frac{1}{k}\{I_m(u'(R_0),u(R_0))_X+I_m\int_{R_0}^r g(t)dt + \int_r^\infty e^{2ik(t-r)}g(t)dt\},$$

whence follows that

$$(7.12)\quad \lim_{r\to\infty}|v(r)|_X^2=\frac{1}{k}\{I_m(u'(R_0),u(R_0))_X+I_m\int_{R_0}^\infty g(t)dt\},$$

which completes the proof.

Q.E.D.

LEMMA 7.2. Let v be as in Lemma 7.1. Then there exists the weak limit

$$(7.13)\quad F(k,f)=w\text{-}\lim_{r\to\infty} e^{-i\mu(r\cdot,k)}v(r)$$

in X, where $\mu(y,k)=rk-\lambda(y)$ and $\lambda(y)$ is given by (5.40).

PROOF. Let us set

$$(7.14)\quad \begin{cases} \alpha_k(r)=\frac{1}{2ik}e^{-i\mu(r\cdot,k)}(v'(r)+ikv(r)) \\ \qquad =\frac{1}{2ik}e^{irk}(u'(r)+iku(r)-iZ(r\cdot)v(r)) \\ \tilde{\alpha}_k(r)=e^{irk}(u'(r)-iku(r)) \\ \qquad =e^{irk+i\lambda(r\cdot)}(v'(r)-ikv(r)+iZ(r\cdot)v(r)) \end{cases}$$

with $u=e^{i\lambda}v$. Let $x\in D$. Then, by (i) of Lemma 6.2,

$$(7.15)\quad \begin{aligned} \frac{d}{dr}(\alpha_k(r),x)_X &= -\frac{1}{2ik}e^{-irk}(-(u'-iku)'-ik(u'-iku)+iZ'u+iZ(u'-iku),x)_X \\ &= \frac{1}{2ik}e^{-irk}\{(u,Bx)_X-(e^{i\lambda}f,x)_X+i(Z(u'-iku),x)_X \\ &\quad +((C_1-iP-Y)u,x)_X-2ir^{-2}(u,Mx)_X\}=\frac{1}{2ik}e^{-irk}g(r,x), \end{aligned}$$

where we have used the relation (6.5) in Lemma 6.3 with $x=u$, $x'=x$ and $S(y)\equiv 1$. Similarly we have

$$(7.16)\quad \begin{aligned} \frac{d}{dr}(\tilde{\alpha}_k(r),x)=e^{irk}\{&(u,Bx)_X-(e^{i\lambda}f,x)+2i(Z(u'-iku),x)_X \\ &-((Y-C_1-iZ'+iP)u,x)_X-2ir^{-2}(u,Mx)_X\}=e^{irk}h(r,x) \end{aligned}$$

Therefore, by (5.10) in Theorem 5.4 and the fact that $|Mx|_X\le Cr^{1-\varepsilon}|A^{\frac{1}{2}}x|_X$, $g(r,x)$ and $h(r,x)$ are integrable over $(1,\infty)$, which implies that there exist limits

$$(7.17) \begin{cases} \lim_{r\to\infty}(\alpha_k(r),x)_x = \alpha(k,x), \\ \lim_{r\to\infty}(\tilde{\alpha}_k(r),x)_x = \tilde{\alpha}(K,x). \end{cases}$$

Here $\tilde{\alpha}(k,x)=0$. In fact, since $u'-iku \in L_{2,\beta}(I,X)$, there exists a sequence $\{r_n\}$ such that $|\tilde{\alpha}_k(r_n)|_x$ tends to zero as $n\to\infty$. Therefore, by taking account of the fact that $(Z(r\cdot)v(r),X)_x \to 0$ as $r\to\infty$, which is obtained from the boundedness of $|v(r)|_x$ ($r\in I$), it follows from the second relation of (7.17) that

$$(7.18) \begin{aligned} &\lim_{r\to\infty} e^{-i\mu(r\cdot,k)}(v'(r)-ikv(r),x)_x \\ &= \lim_{r\to\infty}\{e^{-2irk}(\tilde{\alpha}_k(r),x)_x - ie^{-i\mu(r\cdot,k)}(Z(r\cdot)v(r),x)_x\}=0 \end{aligned}$$

Thus we obtain from (7.18) and the first relation of (7.17)

$$(7.19) \begin{cases} \lim_{r\to\infty}\{(e^{-i\mu(r\cdot,k)}v'(r),x)_x + ik(e^{-i\mu(r\cdot,k)}v(r),x)_x\}=2ik\,\alpha(k,x), \\ \lim_{r\to\infty}\{(e^{-i\mu(r\cdot,k)}v'(r),x)_x - ik(e^{-i\mu(r\cdot,k)}v(r),x)_x\}=0 \end{cases}$$

Whence follows the existence of the limits

$$(7.20) \begin{aligned} \alpha(k,X) &= \lim_{r\to\infty}(e^{-i\mu(r\cdot,k)}v(r),x)_x \\ &= \frac{1}{ik}\lim_{r\to\infty}(e^{-i\mu(r\cdot,k)}v'(r),x)_x \end{aligned}$$

for $x\in D$. Because of the denseness of D in X and the boundedness of $|v(r)|_x$ on I, which completes the proof.

$$\mathcal{Q}.E.D.$$

COROLLARY 7.3. Let $\{R_n\}$ be a sequence such that $R_n\uparrow\infty$ and $|v'(R_n)-ikv(R_n)|\to 0$ as $n\to\infty$. Then there exists the weak limit

$$(7.21)\quad F(k,f) = \text{w-}\lim_{n\to\infty}\alpha_k(R_n) \text{ in } X.$$

PROOF. Since $\{|v'(R_n)|_x\}$ is a bounded sequence as well as $\{|v(R_n)|_x\}$, it follows from (7.20) that the weak limit

$\text{w-}\lim_{n\to\infty} e^{-i\mu(R_n\cdot,k)}v'(R_n)$ exists in X and is equal to $ikF(k,f)$. Therefore, (7.21) is easily obtained from the definition of $\alpha_k(r)$.

Q.E.D.

LEMMA 7.4. Let v be as above. There exists a sequence $\{r_n\}$ such that $r_n \uparrow \infty$ $(n\to\infty)$ and

(7.22) $\lim_{r\to\infty} |v(r_n)|_X = |F(k,f)|_X$.

PROOF. Let us take a sequence $\{r_n\}$ which satisfies

$$(7.23)\begin{cases} (1+r_n)^{\beta+\frac{1}{2}}|B^{\frac{1}{2}}(r_n)u(r_n)|_X \leqq C_0 \\ \lim_{n\to\infty} (1+r_n)^{\beta+\frac{1}{2}}|B^{\frac{1}{2}}(r_n)u(r_n)|_X = 0 \\ |v(r_n)|_X \leqq C_0, \\ \lim_{n\to\infty} |v'(r_n)-ikv(r_n)|_X = 0, \end{cases}$$

for all $n=1,2,\ldots$, where $C_0>0$ is a constant, $u=e^{i\lambda}v$, and β is as above. Such a $\{r_n\}$ surely exists by Theorem 5.4 and Theorem 2.3. Let us set

(7.24) $\alpha_{-k}(r) = -\frac{1}{2ik}e^{i\mu(r\cdot,k)}(v'(r)-ikv(r))$.

Then we have

(7.25) $v(r) = e^{i\mu(r\cdot,k)}\alpha_k(r) + e^{-i\mu(r\cdot,k)}\alpha_{-k}(r)$,

the definition of $\alpha_k(r)$ being given in (7.13), whence follows that

$$(7.26)\begin{cases} |v(r_n)|_X^2 = (\alpha_k(r_n),\ e^{-i\mu(r_n\cdot,k)}v(r_n))_X \\ \qquad\qquad + (\alpha_{-k}(r_n),\ e^{+i\mu(r_n\cdot,k)}v(r_n))_X = a_n + b_n \end{cases}$$

Here $b_n\to 0$ as $n\to\infty$ because of the third and fourth relations of (7.23). Therefore, it suffices to show that $a_n\to|F(k,f)|_X^2$ as $n\to\infty$. Setting

(7.27) $F_n(k,f) = e^{-i\mu(r_n\cdot,k)}v(r_n)$,

we obtain from (7.15)

$$a_n = (\alpha_k(R), F_n(k,f))_X - \int_{r_n}^{R} \frac{d}{dt}(\alpha_k(t), F_n(k,f))_X dt$$

(7.28)

$$= (\alpha_k(R), F_n(k,f))_X - \frac{1}{2ik}\int_{r_n}^{R} e^{-ik(t-r_n)} g(t,u(r_n))dt,$$

where

(7.29) $$g(r,x) = (Bu,x)_X + ((C_1 - iP - Y)u - e^{i\lambda}f, x)_X + i(Z(u'-iku),x)_X + \{-2ir^{-2}(u,Mx)_X\} = \sum_{j=1}^{4} g_j(r,x).$$

Now let us estimate $g(r,u(r_n))$. It follows from (5.5), (5.42) and the third relation of (7.23) that

(7.30) $$|g_2(r,u(r_n))| \leq C_0(|f(r)|_X + C(1+r)^{-2\delta}|v(r)|_X) = g_{02}(r).$$

As for $g_3(r,u(r_n))$ we have similarly

(7.31) $$|g_3(r,u(r_n))| \leq C_0C(1+r)^{-\varepsilon-\beta}|(1+r)^{\beta}(u'-iku)|_X \leq C_0C(1+r)^{-\delta}|(1+r)^{\beta}(u'-iku)|_X = g_{03}(r),$$

because

(7.32) $$-\varepsilon-\beta = \begin{cases} -\varepsilon \leq -\delta & (\beta=0), \\ -\varepsilon-(\delta-\varepsilon) = -\delta & (\beta=\delta-\varepsilon). \end{cases}$$

$g_4(r,u(r_n))$ can be estimated for $r \geq r_n$ (≥ 1) as

$$|g_4(r,u(r_n))| \leq C(1+r)^{-2}(1+r)^{1-\varepsilon}|u|_X|A^{\frac{1}{2}}u(r_n)|_X$$

(7.33)

$$= C(1+r)^{-1-\varepsilon}(1+r_n)^{\frac{1}{2}-\beta}|u|_X|(1+r_n)^{\beta+\frac{1}{2}}B^{\frac{1}{2}}(r_n)u(r_n)|_X$$

$$\leq CC_0(1+r)^{-\varepsilon-\beta-\frac{1}{2}}|v|_X \leq C'(1+r)^{-\frac{1}{2}-\alpha} = g_{04}(r),$$

where $\alpha=\delta$ $(0<\varepsilon\leq\frac{1}{2})$, $= \varepsilon$ $(\frac{1}{2}<\varepsilon\leq 1)$. Here we have used (5.11) in Theorem 5.4, (7.32) and the first relation of (7.23). Let us enter into the estimation of $g(r,u(r_n))$. We have from the inequality $a\beta \leq \frac{1}{2}(a^2+\beta^2)$

$$|g_1(r,u(r_n))| \leq \tfrac{1}{2}|B^{\frac{1}{2}}u|_X^2 + \tfrac{1}{2}r^{-2}|A^{\frac{1}{2}}u(r_n)|_X^2$$

(7.34)

$$\leq \tfrac{1}{2}|(1+r)^{\beta}B^{\frac{1}{2}}u|_X^2 + \tfrac{1}{2}r^{-2}|A^{\frac{1}{2}}u(r_n)|_X^2$$

$$\equiv g_{01}(r) + \tfrac{1}{2}r^{-2}|A^{\frac{1}{2}}u(r_n)|_X^2$$

Let $g_0(r) = \sum_{j=1}^{4} g_{0j}(r)$. Then g_0 is integrable and it follows from (7.28) that

$$(7.35)\quad |a_n-(\alpha_k(R),F_n(k,f))_X| \leq \frac{1}{2|k|}\int_{r_n}^{R} g_0(r)dr+\frac{1}{4|k|}|A^{\frac{1}{2}}u(r_n)|_X^2\int_{r_n}^{R}\frac{dr}{r^2}$$

$$\leq \frac{1}{2|k|}\int_{r_n}^{R} g_0(r)dr+\frac{1}{4|k|}r_n|B^{\frac{1}{2}}(r_n)u(r_n)|_X^2 \qquad (r_n\geq 1).$$

Let $R\to\infty$ in (7.35) along a sequence such that $|v'(R_n)-ikv(R_n)|_X\to 0$ as $n\to\infty$. Then, using Corollary 7.3, we arrive at

$$(7.36)\quad |a_n-(F(k,f),F_n(k,f))_X| \leq \frac{1}{2|k|}\int_{r_n}^{\infty} g_0(r)dr+\frac{1}{4|k|}r_n|B^{\frac{1}{2}}(r_n)u(r_n)|_X^2$$

By the second relation of (7.23) and Lemma 7.2 the right-hand side of (7.36) tends to zero and $(F(k,f),F_n(k,f))_X$ coverages to $|F(k,f)|_X^2$ as $n\to\infty$, which completes the proof.

Q.E.D.

PROOF of THEOREM 5.3. Let v be the radiative function for $\{L,k,\ell\}$ with $k\in \mathbb{R}-\{0\}$ and $\ell\in F_\beta(I,X)$. According to Lemma 2.8 v is rewritten as $v=v_0+w$, where v_0 is the radiative function for $\{L,k_0,\ell\}$ with $k_0\in\mathbf{C}^+$, $I_m k_0>0$ and w is the radiative function for $\{L,k,\ell[f]\}$ with $f=(k^2-k_0^2)v_0$. Since $v_0\in H_0^{1,B}(I,X)$, it can be easily seen that

$$(7.37)\quad s\text{-}\lim_{r\to\infty} v_0(r) = 0$$

In fact, letting $R\to\infty$ in the relation

$$(7.38)\quad |v_0(r)|_X^2 = |v_0(R)|_X^2 - 2\int_r^R \mathrm{Re}(v_0'(t),v_0(t))dt$$

along a sequence $\{R_n\}$ which satisfies $|v_0(R_n)|X\to 0$ $(n\to\infty)$, we have

$$(7.39)\quad |v_0(r)|_X^2 \leq 2\int_r^\infty |v_0'(t)|_X|v_0(t)|_X dr \leq \|v_0\|_{B,(r,\infty)}^2,$$

whence (7.37) follows. As for w we can apply Lemmas 7.1, 7.2 and 7.4 to show

$$(7.40)\begin{cases} w\text{-}\lim_{r\to\infty} e^{-i\mu(r\cdot,k)}w(r) = F(k,f) \text{ in } X, \\ \lim_{r\to\infty}|w(r)|_X = |F(k,f)|_X, \end{cases}$$

which implies the strong convergence of $e^{-i\mu(r\cdot,k)}w(r)$.

Therefore the existence of the strong limit

$$(7.41)\quad \begin{aligned} F(k,\ell) &= \underset{r\to\infty}{\text{s-lim}}\ e^{-i\mu(r\cdot,k)}(v_0(r)+w(r)) \\ &= F(k,f) \qquad\qquad (f = (k^2-k_0^2)v_0) \end{aligned}$$

has been proved completely.

Q.E.D.

§8. Some properties of $\lim_{r\to\infty} e^{-i\mu}v(r)$

In this section we shall investigate some properties of $F(k,\ell) = s - \lim_{r\to\infty} e^{-i\mu(r,k)}v(r)$, whose existence has been proved for the radiative function v for $\{L, k, \ell\}$ with $k \in \mathbf{R} - \{0\}$, $\ell \in F_{1+\beta}(I, X)$. The results obtained in this section will be useful when we develop a spectral representation theory in Chapter III.

LEMMA 8.1. Let Assumption 5.1 be satisfied and let $F(k,\ell)$ be as in Theorem 5.3, i.e.,

$$F(k,\ell) = s - \lim_{r\to\infty} e^{-i\mu(r\cdot,k)}v(r) \qquad \text{in } X, \tag{8.1}$$

where v is the radiative function for $\{L, k, \ell\}$ with $k \in F_{1+\beta}(I,X)$, where β is given by (5.9). Then there exists a constant $C = C(k)$ such that

$$|F(k,\ell)|_X \leqq C|||\ell|||_\delta \ . \tag{8.2}$$

$C(k)$ is bounded when k moves in a compact set in $\mathbf{R} - \{0\}$.

PROOF. As in the proof of Theorem 5.3 given at the end of §7, v is decomposed as $v = v_0+w$. Then, as can be easily seen from the proof of Theorem 5.3,

$$|F(k,\ell)|_X = \lim_{r\to\infty}|w(r)|_X \ . \tag{8.3}$$

Set $g_0 = (k^2 - k_0^2)\, v_0$, k_0 being as in the proof of Theorem 5.3. It follows from (3.32) with $v=w$, $f=g_0$, $k_1=k$, $k_2=0$ that

$$(8.4)\qquad \begin{aligned} |w(r)|_X^2 &\leqq \frac{1}{k^2}\Big\{|w'(r) - ikw(r)|_X^2 - 2k\ \mathrm{Im}\ (g_0,w)_{0,(0,r)}\Big\} \\ &\leqq \frac{1}{k^2}|w'(r) - ikw(r)|_X^2 + \frac{2}{|k|}C\|g_0\|_\delta^2 \\ &\leqq \frac{1}{k^2}|w'(r) - ikw(r)|_X^2 + \frac{2}{|k|}C'|||\ell|||_\delta^2 \end{aligned}$$

with $C = C(k)$ and $C' = C'(k)$. Here (2.7) in Theorem 2.3 and Lemma 2.8 have been used. Let $r\to\infty$ along a sequence $\{r_n\}$ such that $r_n\uparrow\infty$ and $|v'(r_n) - ikv(r_n)|_X \to 0$ as $n\to\infty$. Then we obtain (8.2).

Q.E.D.

Since the radiative function $v(\cdot,k,\ell)$ for $\{L,k,\ell\}$ is linear with respect to ℓ , which follows from the uniqueness of the radiative function, a linear operator $F(k)$ from $F_{1+\beta}(I,X)$ into X is well-defined by

$$(8.5)\qquad F(k)\ell = F(k,\ell) \qquad (\ell\in F_{1+\beta}(I,X),\ \beta=\delta-\varepsilon(0<\varepsilon\leqq\tfrac{1}{2}),\ = 0(\tfrac{1}{2}<\varepsilon\leqq 1)) .$$

In the following lemma the denseness of $F_{1+\beta}(I,X)$ in $F_\delta(I,X)$ will be shown, which, together with Lemma 8.1, enables us to extend $F(k)$ uniquely to a bounded linear operator from $F_\delta(I,X)$ into X.

LEMMA 8.2. Let $0 \leqq \varphi \leqq \beta$. Then $F_\beta(I,X)$ is dense in $F_\varphi(I,X)$.

PROOF. As has been shown before the proof of Lemma 2.7 in §4, $\ell\in F_\varphi(I,X)$ can be regarded as a bounded anti-linear functional on $H^{1,B}_{0,-\varphi}(I,X)$ which is defined as the completion of $C_0^\infty(I,X)$ by the norm

$$(8.6)\qquad \|\phi\|^2_{B,-\varphi} = \int_I (1+r)^{-2\varphi}\{|\phi'(r)|_X^2 + |B^{\frac{1}{2}}(r)\phi(r)|_X^2 + |\phi(r)|_X^2\}dr .$$

By the Riesz theorem there exists $w = w_\ell \in H^{1,B}_{0,-\varphi}(I,X)$ such that

(8.7) $$\langle \ell, v \rangle = (w,v)_{B,-\varphi} \qquad (v \in H^{1,B}_{0,-\varphi}(I,X)) ,$$

where $(\ ,\ )_{B,-\varphi}$ is the inner product of $H^{1,B}_{0,-\varphi}(I,X)$. The norm $|||\ell|||_\varphi$ is equivalent to $\|w\|_{B,-\varphi}$. Let $\psi(t)$ be a real-valued, smooth function on $\mathbf{R}$ such that $\psi(t)=1$ $(t\leqq 0)$, $=0$ $(t\geqq 1)$, and set

(8.8) $$\langle \ell_n, v \rangle = (w_n, v)_{B,-\varphi} \qquad (n=1,2,--)$$

with $w_n(r) = \psi(r-n)w(r)$. Obviously $\ell_n \in F_\beta(I,X)$ and we have $\|w-w_n\|_{B,-\varphi} \to 0$ as $n\to\infty$, which implies that

(8.9) $$|||\ell - \ell_n|||_\varphi \to 0 \qquad (n\to\infty) .$$

Thus the denseness of $F_\beta(I,X)$ in $F_\varphi(I,X)$ $(0\leqq \varphi \leqq \beta)$ has been proved.

Q.E.D.

From Lemmas 8.1 and 8.2 we see that the operator $F(k)$, $k \in \mathbf{R} - \{0\}$, can be extended uniquely to a bounded linear operator from $F_\delta(I,X)$ into X. Thus we give

DEFINITION 8.3. We denote again by $F(k)$ the above bounded linear extension of $F(k)$. When $\ell=\ell[f]$ with $f \in L_{2,\delta}(I,X)$, we shall simply write

(8.10) $$F(k)\ell[f] = F(k)f .$$

Now we shall show that $F(k)\ell$ can be represented by ℓ and the radiative function $v = v(\cdot,k,\ell)$ for $\{L,k,\ell\}$. As we have seen above, $\ell \in F_\delta(I,X)$ can be regarded as a bounded anti-linear functional on $H^{1,B}_{0,-\delta}(I,X)$. On the other hand any radiative function v for $\{L,k,\ell\}$ $(k \in \mathbf{C}^+, \ell \in F_\delta(I,X))$ belongs to $H^{1,B}_{0,-\delta}(I,X)$. Therefore $\langle \ell, v\rangle$ is well-defined, and we have

(8.11) $$\langle \ell, v \rangle = \lim_{n\to\infty} \langle \ell, v_n \rangle ,$$

where $v_n \in H_0^{1,B}(I,X)$ satisfies $v_n \to v$ in $H_{0,-\delta}^{1,B}(I,X)$.

THEOREM 8.4. Let Assumption 5.1 be satisfied. Let $v_j (j=1,2)$ be the radiative function for $\{L,k,\ell_j\}$ with $k \in \mathbf{R} - \{0\}$, $\ell_j \in F_\delta(I,X)$ $(j=1,2)$. Then

$$(8.12) \qquad (F(k)\ell_1, F(k)\ell_2)_X = \frac{1}{2ik}\{\overline{\langle \ell_2, v_1\rangle} - \langle \ell_1, v_2\rangle\} ,$$

where the right-hand side is well-defined as stated above and the bar means the complex conjugate. In particular

$$(8.13) \qquad |F(k)\ell|_X^2 = -\frac{1}{k}\,\mathrm{Im}\,\langle \ell, v\rangle$$

for the radiative function v for $\{L,k,\ell\}$ with $k \in \mathbf{R} - \{0\}$ and $\ell \in F_\delta(I,X)$. When $\ell_j = \ell[f_j]$ with $f_j \in L_{2,\delta}(I,X), j=1,2,$ (8.12) takes the form

$$(8.14) \qquad (F(k)f_1, F(k)f_2)_X = \frac{1}{2ik}\{(v_1,f_2)_0 - (f_1,v_2)_0\} .$$

Further we have

$$(8.15) \qquad |F(k)f|_X^2 = \frac{1}{k}\,\mathrm{Im}(v,f)_0$$

for the radiative function v for $\{L,k,\ell[f]\}$ with $k \in \mathbf{R} - \{0\}$ and $f \in L_{2,\delta}(I,X)$.

PROOF. Let us first consider a special case that $\ell_j \in F_{1+\beta}(I,X)$, $j=1,2$, where β is given by (5.9). Starting with the relation $(v_1, (L-k^2)\phi)_0 = \langle \ell_1, \phi\rangle$ $(\phi \in C_0^\infty(I,X))$, we obtain

$$(8.16) \qquad \int_I \{(v_1', \phi')_X + (B^{\frac{1}{2}}v_1, B^{\frac{1}{2}}\phi)_X + ((C-k^2)v_1,\phi)_X\}dr = \langle \ell_1, \phi\rangle .$$

Let $\rho(r)$ be a real-valued C^1 function on I such that $0 \leqq \rho(r) \leqq 1$, $\rho(r)=1$ $(r \leqq 1)$, $=0 (r \geqq 2)$ and let us set

$$\rho_n(r) = \rho(\tfrac{r}{n}) \qquad (n=1,2,\dots) \ . \tag{8.17}$$

Then it can be easily seen that

$$\begin{cases} \rho_n'(r) = 0 & \text{if } r < n \text{ or } r > 2n \ , \\ |\rho_n'(r)| \leqq \frac{c}{r} & (r \in I, \ \frac{1}{2} c = \max_{r \in I} |\rho'(r)|). \end{cases} \tag{8.18}$$

Substitute $\phi = \rho_n v^2$ in (8.16). Then we have

$$\int_I \Big[\rho_n'(v_1', v_2)_X + \rho_n\{(v_1', v_2')_X + (B^{\frac{1}{2}} v_1, B^{\frac{1}{2}} v_2)_X + ((C-k^2) v_1 \ v_2)_X\} \Big] dr = \langle \ell_1, \rho_n v_2 \rangle \ . \tag{8.19}$$

Quite similarly, by starting with $((L-k^2)\phi, v_2) = \langle \overline{\ell_2, \phi} \rangle$, it follows that

$$\int_I \Big[\rho_n'(v_1, v_2')_X + \rho_n\{(v_1', v_2')_X + (B^{\frac{1}{2}} v_1, B^{\frac{1}{2}} v_2)_X + ((C-k^2) v_1 \ v_2)_X\} \Big] dr = \overline{\langle \ell_2, \rho_n v_1 \rangle} \tag{8.20}$$

(8.19) and (8.20) are combined to give

$$\int_n^{2n} \rho_n'\{(v_1, v_2')_X - (v_1', v_2')_X\} dr = \langle \ell_2, \rho_n v_1 \rangle - \langle \ell_1, \rho_n v_2 \rangle . \tag{8.21}$$

Since

$$\begin{aligned} &(v_1(r), v_2'(r))_X - (v_1'(r), v_2(r))_X \\ &= (v_1(r), v_2'(r) - ikv_2(r))_X - (v_1'(r) - ikv_1(r), v_2(r))_X \\ &\quad - 2ik(v_1(r), v_2(r))_X \end{aligned} \tag{8.22}$$

and

$$(8.23)\qquad (v_1(r),v_2(r))_X = (e^{-i\mu}v_1(r),\ e^{-i\mu}v_2(r))_X = (F(k)\ell_1,\ F(k)\ell_2)_X + h(r)$$

with $h(r)\to 0$ as $r\to\infty$ by Theorem 5.3, we obtain, noting that (8.18) and

$$(8.24)\qquad \int_n^{2n} \rho_n'(r)dr = \Big[\rho_n(r)\Big]_n^{2n} = -1 \quad ,$$

$$(8.25)\qquad \left| \int_n^{2n} \rho_n'\{(v_1'v_2)_X - (v_1,v_2')_X\}dr - 2ik(F(k)\ell_1,F(k)\ell_2)_X \right|$$

$$\leqq c\int_n^{2n}\left\{ \frac{|h(r)|}{r} + (r^{-\delta}|v_1|_X)(r^{\delta-1}|v_2' - ikv_2|_X) + (r^{\delta-1}|v_1' - ikv_1|_X) \times (r^{-\delta}|v_2|_X)\right\} dr$$

$$\leqq c\left\{2\max_{r\geqq n}|h(r)| + \|v_1\|_{-\delta,(n,\infty)}\|v_2' - ikv_2\|_{\delta-1,(n,\infty)} + \|v_1' - ikv_1\|_{\delta-1,(n,\infty)}\|v_2\|_{-\delta,(n,\infty)}\right\}$$

$$\to 0 \qquad (n\to\infty) \quad .$$

On the other hand, $\rho_n v_j$ tends to v_j in $H^{1,B}_{0,-\delta}(I,X)$ as $n\to\infty$, and hence

$$(8.26)\qquad \overline{\langle \ell_2,\rho_n v_1\rangle} - \langle \ell_1,\rho_n v_2\rangle \to \overline{\langle \ell_2,v_1\rangle} - \langle \ell_1,v_2\rangle$$

as $n\to\infty$. (8.12) follows from (8.25) and (8.26). Let us next consider the general case that $\ell_j\in F_\delta(I,X)$. Then there exist sequences $\{\ell_{1n}\}$ and $\{\ell_{2n}\}$ such that

$$(8.27)\qquad \begin{cases} \ell_{jn}\in F_{1+\beta}(I,X) & (n=1,2,\ldots,\ j=1,2)\ , \\ \ell_{jn}\to \ell_j \ \text{ in } \ F_\delta(I,X) & (j=1,2)\ . \end{cases}$$

Let v_{jn} (j=1,2, n=1,2,...) be the radiative function for $\{L,k,\ell_{nj}\}$. Then we have from Theorem 2.3

(8.28) $$v_{jn} \to v_j \qquad \text{in } H^{1,B}_{0,-\delta}(I,X)$$

as $n\to\infty$ for j=1,2. (8.12) in the general case is easily obtained by letting $n\to\infty$ in the relation

(8.29) $$(F(k)\ell_{1n}, F(k)\ell_{2n})_X = \overline{\langle \ell_{2n}, v_{1n} \rangle} - \langle \ell_{1n}, v_{2n} \rangle ,$$

which has been proved already. (8.13) ~ (8.15) directly follows from (8.12).

Q.E.D.

Now it will be shown that the range $\{F(k)f/f \in L_{2,\delta}(I,X)\}$ contains D, and hence it is dense in X . Let $\xi(r)$ be a smooth function on I defined by (5.36), i.e., $0 \leqq \xi(r) \leqq 1$ and $\xi(r)=0$ $(r \leqq 1)$, $= 1 (r \leqq 2)$. Set

(8.30) $$\begin{cases} w_0(r) = \xi(r)e^{i\mu(r\cdot,k)}x , \\ f_0(r) = (L-k^2)w_0 \end{cases}$$

for $x \in D$ and $k \in \mathbf{R} - \{0\}$. Then, as can be easily seen from the definition of $\mu(y,k)$ and (5.20) in Lemma 5.5, $f_0 \in L_{2,\delta}(I,X)$ and w_0 is the radiative function for $\{L,k,\ell[f_0]\}$.

PROPOSITION 8.5. Let $x \in D$ and let f_0 be as above. Then

(8.31) $$F(k)f_0 = x .$$

PROOF. Let $\chi_n(r)$ be the characteristic function of the interval (0,n), n=1,2,..., and let w_n and η_n be the radiative functions for $\{L,k,\ell[\chi_n f_0]\}$ and $\{L,k,\ell[(1-\chi_n)f_0]\}$, respectively. Note that

$\chi_n f_0 \in L_{2,1+\beta}(I,X)$, because the support of $\chi_n f_0$ is compact. Then, by the relation $w_0 = w_n + \eta_n$ and Theorem 5.3, we have

$$(8.32) \qquad F(k)(\chi_n f) = \text{s-}\lim_{r\to\infty} \{e^{-i\mu(r\cdot,k)} w_0(r) - e^{-i\mu(r\cdot,k)}\eta_n(r)\}$$

in X. Since $e^{-i\mu} w_0(r) = x$ for $r \geqq 2$, there exists the limit

$$(8.33) \qquad x_n = \text{s} - \lim_{r\to\infty} e^{-i\mu(r\cdot,k)}\eta_n(r) \qquad \text{in } X,$$

and hence the relation

$$(8.34) \qquad F(k)(\chi_n f) = x - x_n$$

is valid for each $n=1,2,\ldots$. On the other hand, proceeding as in the derivation of (8.4), we obtain from (3.32),

$$(8.35) \qquad |\eta_n(r)|_X^2 \leqq \frac{1}{k^2}|\eta_n'(r) - ik\eta_n(r)|_X^2 + \frac{C}{|k|}\|(1-\chi_n)f_0\|_\delta^2 \qquad (n=1,2,\ldots)$$

with $C=C(k)$, whence follows by letting $r\to\infty$ along a suitable sequence $\{r_m\}$ in (8.32) that

$$(8.36) \qquad |x_n|_X^2 \leqq \frac{C}{|k|}\|(1-\chi_n)f_0\|_\delta^2 \qquad (n-1,2,\ldots) .$$

Thus we arrive at

$$(8.37) \qquad |F(k)(\chi_n f_0) - x|_X \leqq C\|(1-\chi_n)f_0\|_\delta \qquad (C = C(k),\ n=1,2,\ldots) .$$

When n tends to infinity, $\chi_n f_0 \to f_0$, i.e., $(1-\chi_n)f_0 \to 0$ in $L_{2,\delta}(I,X)$, and (8.31) is obtained from (8.37).

Q.E.D.

In the remainder of this section we shall consider the restriction of $F(k)$ to $L_{2,\delta}(I,X)$, which is denoted by $F(k)$ again. Let us consider the mapping

$$\mathbb{C}^{+} \ni k \to F(k) \in \mathbf{B}(L_{2,\delta}(I,X),X) \quad . \tag{8.38}$$

By the use of Lemma 4.3 $F(k)$ can be seen a $\mathbf{B}(L_{2,\delta}(I,X),X)$ - valued continuous function on $\mathbf{R}-\{0\}$. Further we can see that $F(k)$ is a compact operator from $L_{2,\delta}(I,X)$ into X.

THEOREM 8.6. Let Assumption 5.1 be satisfied.

(I) Then $F(k)$ is a $\mathbf{B}(L_{2,\delta}(I,X),X)$ -valued continuous function on $\mathbf{R}-\{0\}$.

(II) For each $k \in \mathbf{R}-\{0\}$ $F(k)$ is a compact operator from $L_{2,\delta}(I,X)$ into X.

PROOF. Let us assume that (i) is not true at a point $k \in \mathbf{R}-\{0\}$. Then there exist a positive number $c > 0$ and sequences $\{f_n\} \subset L_{2,\delta}(I,X)$, $\{k_n\} \subset \mathbf{R}-\{0\}$ such that

$$\begin{cases} \|f_n\|_\delta = 1 & (n=1,2,\dots) \\ k_n \to k & (n \to \infty) \\ |F(k_n)f_n - F(k)f_n|_X \geqq c & (n=1,2,\dots) \quad . \end{cases} \tag{8.39}$$

With no loss of generality $\{f_n\}$ may be assumed to converge weakly in $L_{2,\delta}(I,X)$ to some $f \in L_{2,\delta}(I,X)$. We shall show that there exists subsequence $\{n_m\}$ of positive integers such that $\{F(k)f_{n_m}\}$ and $\{F(k_{n_m})f_{n_m}\}$ converge strongly to the same limit $F(k)f$ as $m \to \infty$, which will contradicts the third relation of (8.39). We shall consider the sequence $\{F(k_n)f_n\}$ only, because the sequence $\{F(k)f_n\}$ can be treated in quite a similar way. It follows from (8.13) with $k=k_n$, $f=f_n$, $v=v_n$ that

$$(8.40)\qquad |F(k_n)f_n|_X^2 = \frac{1}{k_n}\mathrm{Im}(v_n,f_n)_0 \qquad (n=1,2\dots)$$

where v_n is the radiative function for $\{L,k_n,\ell[f_n]\}$. By Lemma 4.3 a subsequence $\{v_{n_m}\}$ of $\{v_n\}$ can be chosen to satisfy $\{v_{n_m}\}$ converges in $L_{2,-\delta}(I,X)$ to the radiative function v for $\{L,k,\ell[f]\}$, whence we obtain

$$(8.41)\qquad \lim_{m\to\infty}|F(k_{n_m})f_{n_m}|_X^2 = \frac{1}{k}\mathrm{Im}(v,f) = |F(k)f|_X^2 .$$

Let $x\in D$ and set

$$(8.42)\qquad \begin{cases} w_{n_m}(r) = \xi(r)e^{i\mu(r\cdot,k_{n_m})}x, \\ g_{n_m}(r) = (L-k_{n_m}^2)w_{n_m}(r) , \end{cases}$$

where $\xi(r)$ is defined by (5.36). Then it follows from Proposition 8.5 that

$$(8.43)\qquad F(k_{n_m})g_{n_m} = x .$$

By taking note of Remark 5.9, (2) it can be easily seen that

$$(8.44)\qquad \begin{cases} w_{n_m}\to w_0 & \text{in } L_{2,-\delta}(I,X) \\ g_{n_m}\to f_0 & \text{in } L_{2,\delta}(I,X) \end{cases}$$

as $m\to\infty$ with w_0 and f_0 defined by (8.30). Therefore we have, using (8.13) in Theorem 8.4,

$$\lim_{m\to\infty}(F(k_{n_m})f_{n_m},x)_X = \lim_{m\to\infty}(F(k_{n_m})f_{n_m},\ (k_{n_m})g_{n_m})_X$$

$$= \lim_{m\to\infty}\left|\frac{1}{2ik_{n_m}}\left\{(v_{n_m},g_{n_m})_0 - (f_{n_m},w_{n_m})_0\right\}\right| \tag{8.45}$$

$$= \frac{1}{2ik}\{(v,f_0)_0 - (f,w_0)_0\}$$

$$= (F(k)f,\ F(k)f_0)_X = (F(k)f,\ x)_X ,$$

which, together with (8.38) and the denseness of D in X, implies that $\{F(k_{n_m})f_{n_m}\}$ converges to $F(k)f$ strongly in X. Thus we have shown (i). In order to prove (ii) it is sufficient to show that $\{F(k)f_n\}$ is relatively compact when $\{f_n\}$ is a bounded sequence in $L_{2,\delta}(I,X)$. This can be shown in quite a similar way used in the proof of (i).

Q.E.D.

THEOREM 8.7. Let Assumption 5.1 be satisfied. Let $f\in L_{2,\delta}(I,X)$ and let $k\in\mathbf{R}-\{0\}$. Then

$$F(k)f = s - \lim_{n\to\infty} e^{-i\mu(r_n\cdot,k)}v(r_n) \qquad \text{in } X, \tag{8.46}$$

where v is the radiative function for $\{L,k,\ell[f]\}$ and $\{r_n\}$ is an sequence such that $r_n\uparrow\infty$ and $v'(r_n) - ikv(r_n)\to 0$ in X as $n\to\infty$.

Before showing this theorem we need the following the Green formula.

LEMMA 8.8. Let $v_j \in L_2(I,X)_{loc}$ satisfy the equation

$$(8.47) \qquad (v_j\ (L - \overline{k}^2)\ \phi)_0 = (f_j, \phi)_0 \quad (\phi \in C_0^\infty(I,X),\ j=1,2)$$

with $f_j \in L_2(I,X)_{loc}$ and $k \in \mathbf{R}-\{0\}$. Then we have

$$(8.48) \qquad \begin{aligned} &\int_0^r \{(v_1, f_2)_X - (f_1, v_2)_X\}\, dr \\ &\quad = (v_1'(r) - ikv_1(r),\ v_2(r))_X - (v_1(r), v_2'(r) - ikv_2(r))_X \\ &\qquad\qquad + 2ik\ (v_1(r),\ v_2(r))_X \end{aligned}$$

for $r \in I$.

PROOF. The idea of the proof resembles the one of the proof of Proposition 3.4. As has be shown in the proof of Proposition 1.3, $\tilde{v}_j = U^{-1} v_j \in H_2(\mathbf{R}^N)_{loc}$ $(j=1,2)$. Therefore we can proceed as in the proof of Proposition 3.4 to show that there exist sequences $\{\phi_{1n}\}$ and $\{\phi_{2n}\}$ such that $\phi_{1n}, \phi_{2n} \in C_0^\infty(I,X)$ and

$$(8.49) \qquad \begin{cases} \phi_{jn} \to v_j & \text{in } L_2(I,X)_{loc} \\ \phi_{jn}(r) \to v_j(r) & \text{in } X \quad (r \in \overline{I}) \\ \phi_{jn}'(r) \to v_j'(r) & \text{in } X \quad (r \in I) \\ (L-k^2)\phi_{jn} = f_{nj} \to f_j & \text{in } L_2(I,X)_{loc} \end{cases}$$

as $n \to \infty$ $(j=1,2)$. Integrate the relations

$$(8.50) \qquad \begin{cases} ((L-k^2)\phi_{1n}, \phi_{2n})_X = (f_{1n}, \phi_{2n})_X \\ (\phi_{1n}, (L-k^2)\phi_{2n})_X = (\phi_{1n}, f_{2n})_X \end{cases}$$

from 0 to r and make use of partial integration. Then it follows that

$$(8.51)\qquad \begin{aligned} -(\phi_{1n}'(r),\ \phi_{2n}(r))_X + (\phi_{1n}(r),\phi'_{2n}(r))_X &= (f_{1n},\phi_{2n})_{0,(0,r)} \\ &\quad - (\phi_{1n}, f_{2n})_{0,(0,r)} \end{aligned}$$

where we should note that $(\phi_{1n}'(0),\phi_{2n}(0))_X = (\phi_{1n}(0),\phi_{2n}'(0)) = 0$. (8.48) is easily obtained by letting $n\to\infty$ in (8.51).

Q.E.D.

<u>PROOF of THEOREM 8.7.</u> Set in (3.32) $k_1=k$, $k_2=0$ and $r = r_n$. Then we have

$$(8.52)\qquad |v(r_n)|_X^2 \leqq \frac{1}{k^2}\,|v'(r_n) - iv(r_n)|_X^2 - \frac{2}{k}\mathrm{Im}(f,v)_{0,(0,r_n)}\ ,$$

which implies that $\{|v(r_n)|_X\}$ is a bounded sequence. Next set in (8.48) $v_1 = v_2 = v$, $f_1 = f_2 = f$ and $r = r_n$. Then it follows that

$$(8.53)\qquad \begin{aligned} |e^{-i\mu(r_n\cdot,k)}v(r_n)|_X &= \frac{1}{k}\ \mathrm{Im}(v,f)_{0,(0,r_n)} \\ &\quad + \frac{1}{k}\ \mathrm{Im}(v(r_n),\ v'(r_n) - ikv(r_n))_X \end{aligned}$$

The second term of the right-hand side of (8.53) tends to zero as $n\to\infty$ because $|v'(r_n) - ikv(r_n)|_X\to 0$ as $n\to\infty$ and $|v(r_n)|_X$ is bounded, and hence we have

$$(8.54)\qquad \lim_{n\to\infty} |e^{-i\mu(r_n\cdot,k)}v(r_n)|_X^2 = \frac{1}{k}\ \mathrm{Im}(v,f)_0 = |F(k)f|_X^2\ ,$$

where (8.15) has been used. Let $x \in D$ and let w_0 and f_0 as in (8.30). Then, setting in (8.48) $v_1=v$, $v_2=w_0$, $f_1=f$, $f_2=f_0$ and $r=r_n$, we obtain

$$(8.55)\quad \begin{aligned} &(e^{-i\mu(r_n\cdot,k)}v(r_n),\ e^{-i\mu(r_n\cdot,k)}w_0(r_n))_X \\ &= \frac{1}{2ik}\{(v(r_n),w_0'(r_n) - ikw_0(r_n))_X - (v'(r_n)-ikv(r_n),\ w_0(r_n))_X \\ &\qquad + (v,f_0)_{0,(0,r_n)} - (f,w_0)_{0,(0,r_n)}\} \ . \end{aligned}$$

Since $e^{-i\mu(r_n\cdot,k)}w_0(r_n) = x$ $(r_n \geqq 2)$ and $w_0'(r_n) - ikw_0(r_n)$ $= -iZ(r_n\cdot)w_0(r_n) \to 0$ in X $(r_n\to\infty)$, it follows from (8.55) that

$$(8.56)\quad \begin{aligned} \lim_{n\to\infty}(e^{-i\mu(r_n\cdot,k)}v(r_n),\ x)_X &= \frac{1}{2ik}\{(v,f_0)_0 - (f,w_0)_0\} \\ &= (F(k)f,\ F(k)f_0)_X = (F(k)f,x)_X \ , \end{aligned}$$

where (8.14) and Proposition 8.5 have been used. From (8.54), (8.56) and the denseness of D (8.46) is seen to be valid.

Q.E.D.

CHAPTER III

SPECTRAL REPRESENTATION

§9. The Green kernel

In this chapter the results obtained in the previous chapters will be combined to develop a spectral representation theory for the operator

$$(9.1) \qquad L = -\frac{d^2}{dr^2} + B(r) + C(r)$$

given by (0.20) and (0.21) . Throughout this section the potential $Q(y)$ is assumed to satisfy Assumption 2.1.

Now we shall define the Green kernel $G(r,s,k)$ $(r,s \in \bar{I} = [0,\infty)$, $k \in \mathbb{C}^+)$ and investigate its properties. Let $s \in \bar{I}$, $x \in X$ and let $\ell[s,x]$ be an anti-linear function on $H_0^{1,B}(I,X)$ defined by

$$(9.2) \qquad <\ell[s,x],\phi> = (x,\phi(s))_X \qquad (\phi \in H_0^{1,B}(I,X)) .$$

Then it follows from the estimate

$$(9.3) \qquad |\phi(S)|_X \leqq \sqrt{2}\, \|\phi\|_B \qquad (\phi \in H_0^{1,B}(I,X)) ,$$

which is shown in Proposition 1.1, that $\ell[s,x] \in F_\beta(I,X)$ for any $\beta \geqq 0$ and the estimate

$$(9.4) \qquad |||\, \ell[s,x]\, |||_\beta \leqq \sqrt{2}\, (1 + s)^\beta |x|_X$$

is valid. Denote by $v = v(\cdot,k,s,x)$ the radiative function for $\{L,k,\ell[s,x]\}$. Then, by the use of (1.8) in Proposition 1.3 and (2.7) in Theorem 2.3, we can easily show that

$$(9.5) \qquad |v(r)|_X \leqq C(1 + s)^\delta |x|_X \qquad (C = C(R,k),\ r \in [0,R],\ s \in \bar{I} ,\ x \in X)$$

for any $R \in I$. Therefore a bounded operator $G(r,s,k)$ on X is well-defined by

$$G(r,s,k)x = v(r,k,s,x) . \tag{9.6}$$

DEFINITION 9.1. (the Green kernel). The bounded operator $G(r,s,k)(r,s \in \overline{I}\ ,\ k \in \mathbb{C}^+)$ will be called the Green kernel for L .

The linearity of the operator $G(r,s,k)$ directly follows from the linearity of $\ell[s,x]$ with respect to x . Roughly speaking, $G(r,s,k)$ satisfies

$$(L - k^2)\ G(r,s,k) = \delta(r - s)\ , \tag{9.7}$$

the right-hand side denoting δ-function. The following properties of the Green kernel $G(r,s,k)$ will be made use of further on.

PROPOSITION 9.2. Let Assumption 2.1 be satisfied.

(i) Then $G(\cdot,s,k)x$ is an $L_{2,-\delta}(I,X)$-valued continuous function on $\overline{I} \times \mathbb{C}^+ \times X$. Further, $G(r,s,k)x$ is an X-valued function on $\overline{I} \times \overline{I} \times \mathbb{C}^+ \times X$, too.

(ii) $G(0,r,k) = G(r,0,k) = 0$ for any pair $(r,k) \in I \times \mathbb{C}^+$.

(iii) Let $(s,k,x) \in \overline{I} \times \mathbb{C}^+ \times X$ and let J be an arbitrary open interval such that the closure of J contained in $I - \{s\}$. Then we have $G(\cdot,s,k)x \in D(J)$, where the definition of $D(J)$ is given after the proof of Proposition 1.3.

(iv) Let $R > 0$ and let K be a compact set of $\mathbb{C}^+$. Then there exists $C = C(R,K)$ such that

$$\| G(r,s,k)\| \leqq C \qquad (0 \leqq r,\ s \leqq R,\ k \in K)\ , \tag{9.8}$$

where $\|\ \|$ means the operator norm.

(v) We have for any triple $(r,s,k) \in \overline{I} \times \overline{I} \times \mathbb{C}^+$

$$G(r,s,k)^* = G(s,r,-\overline{k}) , \tag{9.9}$$

$G(r,s,k)^*$ denoting the adjoint of $G(r,s,k)$.

PROOF. Let us first show the continuity of $\ell[s,x]$ in $F_\beta(I,X)$, $\beta \geqq 0$, with respect to $s \in \overline{I}$ and $x \in X$. In fact we obtain from Proposition 1.1.

$$\begin{aligned} |<\ell[s,x] - \ell[s',x'], (1+r)^\beta\phi>| &= |(x, (1+s)^\beta\phi(s)_X - (x', (1+s')^\beta\phi(s')) \\ &\leqq |(x - x', (1+s)^\beta\phi(s))_X| + (x', (1+s)^\beta(\phi(s) - \phi(s')))_X| \\ &\qquad + |(x', ((1+s)^\beta - (1+s')^\beta)\phi(s'))_X| \\ &\leqq \{\sqrt{2}|x - x'|_X(1+s)^\beta + |x'|_X(1+s)^\beta|s-s'|^{\frac{1}{2}} + \sqrt{2}|x'|_X|(1+s)^\beta - (1+s')^\beta|\}\|\phi\|_\beta \end{aligned} \tag{9.10}$$

whence follows the continuity. By recalling that the radiative function v for $\{L,k,\ell\}$ is continuous both in $L_{2,-\delta}(I,X)$ and in $H^{1,B}_0(I,X)_{loc}$ with respect to $R \in \mathbb{C}^+$ and $\ell \in F_\delta(I,X)$, (i) follows immediately. Since $G(\cdot,s,k)x \in H^{1,B}_0(I,X)_{loc}$, we have $G(0,s,k)x = 0$ for all (s,k,x) $\overline{I} \times \mathbb{C}^+ \times X$, which means that $G(0,s,k) = 0$. On the other hand, $<\ell[0,x],\phi> = (x,\phi(0))_X = 0$ for all $\phi \in H^{1,B}_0(I,X)$, and hence $G(\cdot,0,k)x$ is the radiative function for $\{L,k,0\}$. Therefore $G(r,0,k)x \equiv 0$ by the uniqueness of the radiative function. Thus (ii) is completely proved. Let us show (iii). Take a real-valued, smooth function $\psi(r)$ on I such that $\psi(r) = 1$ on J and $\psi(r) = 0$ in a sufficiently small neighborhood of $r = s$. Then it is easy to see that $\hat{v}(r) = \psi(r)G(r,s,k)x$ is the radiative function for $\{L,k,\ell[g]\}$ with $g(r) = -\psi''(r)G(r,s,k)x - 2\psi'(r)\frac{d}{dr} G(r,s,k)x$. Proposition 1.3 can be made use of to show that $\hat{v} \in D(I)$, which means that $G(\cdot,s,k)x \in D(J)$.

(iv) is obvious from (9.5). Let us enter into the proof of (v). Set $v(t) = G(t,s,k)x$ and $w(t) = G(t,r,-\bar{k})x$ with $x,x' \in X$. Then, setting $\phi = \varphi_n w$ and $\varphi_n v$ in the relations

$$(9.11)\begin{cases}(v',\phi')_0 + (B^{\frac{1}{2}}v, B^{\frac{1}{2}}\phi)_0 + (v,(C-\bar{k})\phi)_0 = (x,\phi(s))_X , \\ (\phi',w')_0 + (B^{\frac{1}{2}}\phi, B^{\frac{1}{2}}w)_0 + ((C-k^2)\phi, w)_0 = (\phi(r), x')_X ,\end{cases}$$

respectively, and combining them, we arrive at

$$(9.12)\quad (x,w(s))_X - (v(r),x')_X = \int_0^\infty \varphi_n'(t)\{(v'(t),w(t))_X - (v(t),w'(t))_X\}dt \qquad (n = 1,2,\dots) ,$$

where $\varphi_n(t) = \varphi(n(t-t_0))$, $\varphi(t)$ is a real-valued, smooth function on $\mathbb{R}$ such that $\varphi(t) = 1$ $(t \leqslant 0)$, $= 0$ $(t \geqslant 1)$, and t_0 is taken to satisfy $t_0 > \max(r,s)$. Let $n \to \infty$ in (9.12) and take note of the fact that $\varphi'_n(t) \to -\delta(t-t_0)$ as $n \to \infty$. Then we have

$$(9.13)\quad \begin{aligned}&(x,G(s,r,-\bar{k})x')_X - (G(r,s,k)x,x')_X \\ &= (v(t_0), w'(t_0) + i\bar{k}w(t_0))_X - (v'(t_0) - ikv(t_0), w(t_0))_X \end{aligned} \qquad (t_0 > \max(r-s)) .$$

By making $t_0 \to \infty$ along a suitable sequence $\{t_n\}$, the right-hand side of (9.13) tends to zero, whence follows that

$$(9.14)\qquad (x,\{G(s,r,-\bar{k}) - G(r,s,k)^*\}x')_X = 0$$

for any $x,x' \in X$. Thus we have proved (v).

Q.E.D.

§10. The eigenoperators

The purpose of this section is to construct the eigenoperator $\eta(r,k)$ $(r\in\overline{I}, k\in\mathbf{R}-\{0\})$ by the use of the Green kernel $G(r,s,k)$ which was defined in §9. In this and the following sections $\mathcal{Q}(y)$ will be assumed to satisfy Assumption 5.1 which enables us to apply the results of Chapter II.

We shall first show some more properties of the Green kernel in addition to Proposition 9.2.

<u>PROPOSITION 10.1</u>. Let Assumption 5.1 be satisfied. Then the following estimate for the Green kernel

$$(10.1)\qquad \|G(r,s,k)\| \leqq \begin{cases} C_1(k) & (k\in\mathbf{C}^+, \operatorname{Im}k>0, r,s\in\overline{I}) \\ C_2(k)\min\{(1+r)^{1+\beta}, (1+s)^{1+\beta}\} & \\ & (k\in\mathbf{R}-\{0\}, r,s\in\overline{I}) \end{cases}$$

holds, where $\beta = \delta - \varepsilon$ $(0<\varepsilon\leqq\frac{1}{2})$, $= 0$ $(\varepsilon<\frac{1}{2}\leqq 1)$ and the constants $C_1(k)$ and $C_2(k)$ are bounded when k moves in a compact set in $\{k\in\mathbf{C}^+/\operatorname{Im}k>0\}$ and $\mathbf{R}-\{0\}$, respectively. Further

$$(10.2)\qquad v(r,k,\ell[f]) = \int_I G(r,s,k)f(s)ds \qquad (r\in\overline{I})$$

holds for any radiative function $v(\cdot,k,\ell[f])$ for $\{L,k,\ell[f]\}$ with $k\in\mathbf{C}^+$ and $f\in L_{2,\delta}(I,X)$.

<u>PROOF</u>. (i) Assume that $k\in\mathbf{C}^+$ with $\operatorname{Im}k>0$. Then it follows from Lemma 2.7 that $v = G(,s,k)x$ (s $\overline{I}$, x X) belongs to $H_0^{1,B}(I,X)$. The first estimate of (10.1) is obtained from (9.3), (2.16) in Lemma 2.7 and (9.4) with $\beta=0$. Next assume that $k\in\mathbf{R}-\{0\}$. Applying (5.11)

in Theorem (5.4) and using (9.4), we have $|G(r,s,k)x|_X \leq C(k)(1+S)^{1+\beta}|x|_X$, which implies that

(10.3) $$\|G(r,s,k)\| < C(1+s)^{1+\beta} \qquad (r,s\in\bar{I}) .$$

The second estimate of (10.1) follows from (10.3) and the relation $G(r,s,k)^* = G(s,r,-k)$ ((v) of Proposition 9.2). (ii) Let $f\in L_2(I,X)$ with compact support in $\bar{I}$ and let $k\in\mathbb{C}^+$, $\mathrm{Im}k>0$. Then $u = G(\cdot,r,-\bar{k})x$ and the radiative function v for $\{L,k,\ell[f]\}$ belong to $H_0^{1,B}(I,X)$ and satisfy

(10.4) $$\begin{cases} (u',\phi')_0 + (B^{\frac12}u,B^{\frac12}\phi)_0 + ((C-\bar{k})u,\phi)_0 = (x,\phi(r))_X , \\ (\phi',v')_0 + (B^{\frac12}\phi,B^{\frac12}v)_0 + ((C-\bar{k}^2)\phi,v)_0 = (\phi,f)_0 \end{cases}$$

for $\phi\in H_0^{1,B}(I,X)$. Set $\phi = v$ in the first relation of (10.4) . Then, using the relation $G(s,r,-\bar{k})^*= G(r,s,k)$, we have

(10.5) $$\begin{aligned} (x,v(r))_X &= (u',v')_0 + (B^{\frac12}u,B^{\frac12}v)_0 + ((C - \bar{k}^2)u,v)_0 \\ &= (G(\cdot,r,-k)x,f)_0 = (x,\textstyle\int_I G(r,s,k)f(s)ds)_X \end{aligned}$$

whence (10.2) follows. Let f be as above and $k \in \mathbf{R} - \{0\}$. Then we can approximate k by $\{k + \frac{i}{n}\}$ $(n=1,2,\dots)$ to obtain (10.2), where we have made use of the continuity of the radiative function $v(\cdot,k,\ell[f])$ with respect to k and (9.8) in Proposition 9.2. Thus (10.2) has been established for $k \in \mathbb{C}^+$ and $f \in L_2(I,X)$ with compact support in $\bar{I}$. In the case that $f \in L_{2,\delta}(I,X)$ we can approximate f by $\{f_n\}$, where $f_n\in L_2(I,X)$ with compact support in $\bar{I}$. Then, taking note of the continuity of the radiative function $v(\cdot,k,\ell[f])$ with respect to f and the estimate (10.1), we arrive at (10.2).

Q.E.D.

Let $k \in \mathbf{R} - \{0\}$, $s \in \bar{I}$, $x \in X$ and set $v(r) = G(r,s,k)x$. v is the radiative function for $\{L,k,\ell[s,x]\}$ and $\ell[s,x] \in F_{1+\beta}(I,X)$ with $|||\ell[s,x]|||_{1+\beta} \leqq \sqrt{2}\,(1+s)^{1+\beta}|x|_X$, where β is given by (5.9). Therefore Theorem 5.3 can be applied to show the existence of the limit

$$(10.6) \qquad F(k)\ell[s,x] = s - \lim_{r\to\infty} e^{-i\mu(r\cdot,k)}v(r) \qquad \text{in } X .$$

It follows from Lemma 8.1 that

$$(10.7) \qquad |F(k)\ell[s,x]| \leqq C\|\ell[s,x]\|_\delta = \sqrt{2}\,C(1+s)^\delta|x|_X$$

with $C = C(k)$, and hence for each pair $(r,k) \in \bar{I} \times (\mathbf{R} - \{0\})$ a bounded linear operator $\eta(r,k)$ on X is well-defined by

$$(10.8) \qquad s - \lim_{r\to\infty} e^{-i\mu(t\cdot,k)}G(t,r,k)x = \eta(r,k)x \qquad (x \in X) .$$

<u>DEFINITION 10.2</u>. The bounded linear operator $\eta(r,k)$ defined by (10.8) will be called the <u>eigenoperator</u> associated with L.

The appropriateness of this naming will be justified in the remainder of this section (especially in Theorem 10.4).

<u>PROPOSITION 10.3</u>. Let Assumption 5.1 be satisfied. Then we have

$$(10.9) \qquad s - \lim_{s\to\infty} G(r,s,-k)e^{i\mu(s\cdot,k)}x = \eta^*(r,k)x$$

for any triple $(r,k,x) \in \bar{I} \times (\mathbf{R} - \{0\})\times X$, where $\mu(y,k)$ is defined by (5.8) and $\eta^*(r,k)$ is the adjoint of $\eta(r,k)$.

<u>PROOF</u>. Suppose that there exist $r > 0$, $k_0 \in \mathbf{R} - \{0\}$, $x_0 \in X$, $\beta_0 > 0$ and a sequence $\{s_n\}$ such that $s_n \uparrow \infty$ and $|v_n(r_0) - \eta^*(r_0,k_0)x_0|_X \geqq \beta_0$ holds for all $n=1,2,\ldots,$ where we set $v_n(r) = G(r,s,-k_0)e^{i\mu(s_n\cdot,k)}x_0$. By the use of the interior estimate

(Proposition 1.2) and (10.1) it can be seen that the sequence $\{\|v_n\|_{B,(0,R)}\}$ is bounded for each $R>0$. By Theorem 1.5 (the Rellich theorem) there exists a subsequence $\{w_p\}$ of $\{v_n\}$ such that $\{w_p\}$ is a Cauchy sequence of $L_2(I,X)_{loc}$. Make use of the interior estimate again. Then $\{w_p\}$ is seen to be a Caucy sequence in $H_0^{1,B}(I,X)_{loc}$. Therefore it follows from (1.8) that $\{w_p(r_0)\}$ is a Cauchy sequence in X. Since $G(r_0,s,-k_0)e^{i\mu(s\cdot,k)}x_0$ converges to $\eta^*(r_0,k_0)x_0$ weakly in X by (10.8), $w_p(r_0)$ converges to $\eta^*(r_0,k_0)x_0$ strongly in X, which is a contradiction.

Q.E.D.

Let us summarize these results in the following

THEOREM 10.4. Let Assumption 5.1 be satisfied.

(i) Then

$$\eta(r,k)x = s - \lim_{s\to\infty} e^{-i\mu(s\cdot,k)}G(s,r,k)x \quad \text{in } X, \tag{10.10}$$

and

$$\eta^*(r,k)x = s - \lim_{s\to\infty} G(r,s,-k)e^{i\mu(s\cdot,k)}x \quad \text{in } X \tag{10.11}$$

for any triple $(r,k,x) \in \bar{I} \times (\mathbf{R} - \{0\})\times X$. We have

$$\|\eta(r,k)\| = \|\eta^*(r,k)\| \leqq C(1+r)^{\delta}, \tag{10.12}$$

where $\|\ \ \|$ means the operator norm and $C = C(k)$ is bounded when k moves in a compact set in $\mathbf{R} - \{0\}$.

(ii) The relation

$$2ik(\eta(s,k)x, = (\{G(r,s,k) - G(r,s,-k)\}x,x')_X \tag{10.13}$$

holds for any $x, x' \in X$, $k \in \mathbf{R} - \{0\}$ and $r, s \in \overline{I}$.

(iii) $\eta^*(\cdot,k)x \in H_0^{1,B}(I,X)_{loc} \cap D(I)$ and satifies the equation

$$(L - k^2)v(r) = 0 \tag{10.14}$$

where $r \in \overline{I}$, $k \in \mathbf{R} - \{0\}$, $x \in X$.

PROOF. (i) follows from (10.8), Proposition 11.3 and (10.7). Set in (8.12) $\ell_1 = \ell[s,x]$, $\ell_2 = \ell[r,x']$, $v_1 = G(\cdot,s,k)x$ and $v_2 = G(\cdot,r,k)x$. Then, noting that $F(k)\ell[s,x] = \eta(s,k)x$ and $F(k)\ell[r,x] = \eta(r,k)x'$, we obtain,

$$\begin{aligned}(\eta(s,k)x\,,\,\eta(r,k)x)_X &= \frac{1}{2ik}\Big\{\overline{\langle \ell[r,x'],\,G(\cdot,s,k)x\rangle} \\ &\qquad - \langle \ell[s,x],\,G(\cdot,r,k)x'\rangle\Big\} \\ &= \frac{1}{2ik}\Big\{(G(r,s,k)x,x')_X - (x,G(s,r,k)x')_X\Big\}\end{aligned} \tag{10.15}$$

which, together with the relation $G(s,r,k)^* = G(r,s,-k)$, completes the proof of (ii). Let us show (iii). Set $v_s(r) = G(r,s,-k)e^{i\mu(s\cdot,k)}x$. $v_s(r)$ satisfies

$$(v_s,\,(L - k^2)\phi)_0 = (e^{i\mu(s\cdot,k)}x,\phi(s))_X \qquad (\phi \in C_0^\infty(I,X)). \tag{10.16}$$

Let $\{s_n\}$, $s_n \in I$, be an arbitrary sequence such that $s_n \uparrow \infty$ as $n \to \infty$. Then, as we have seen in the proof of Proposition 10.3, there exists a subsequence $\{t_p\}$ of $\{s_n\}$ such that $\{v_{t_p}\}$ is a Cauchy sequence in $L_2(I,X)$. This means $\{v_s\}$ itself converges to $\eta^*(\cdot,k)x$ in $L_2(I,X)_{loc}$. Letting $s \to \infty$ in (10.16), we arrive at

$$(\eta^*(\cdot,k)x,\,(L - k^2)\phi)_0 = 0 \qquad (\phi \in C_0^\infty(I,X)). \tag{10.17}$$

Proposition 1.3 can be applied to show that $\eta^*(\cdot,k)x \in H_0^{1,B}(I,X)_{loc} \cap D(I)$.

Q.E.D.

The following two theorems will show some relations between the eigenoperator and the operator $F(k)$.

THEOREM 10.5. Let Assumption 5.1 be satisfied. Let $F(k)$, $k \in \mathbf{R} - \{0\}$, be as in §8. Then we have

$$F(k)f = \text{s-}\lim_{R\to\infty} \int_0^R \eta(r,k)f(r)dr \qquad \text{in } X \tag{10.18}$$

for any $f \in L_{2,\delta}(I,X)$. In particular

$$F(k)f = \int_I \eta(r,k)f(r)dr \tag{10.19}$$

holds for $f \in L_{2,\beta}(I,X)$ with $\beta > \frac{1}{2} + \delta$, where the integral is absolutely convergent.

PROOF. Let $f \in L_{2,\delta}(I,X)$ and define $f_R(r)$ by

$$f_R(r) = \begin{cases} f(r) & (0 \leqq r \leqq R), \\ 0 & (r > R). \end{cases} \tag{10.20}$$

Then it follows from Theorem 5.3 and (10.2) that

$$\begin{aligned} F(k)f_R &= \text{s - }\lim_{r\to\infty} e^{-i\mu(r\cdot,k)} v(r,k,\ell[f_R]) \\ &= \text{s - }\lim_{r\to\infty} e^{-i\mu(r\cdot,k)} \int_I G(r,s,k)f_R(s)ds \\ &= \int_0^R \text{s - }\lim_{r\to\infty} \left\{ e^{-i\mu(r\cdot,k)} G(r,s,k)f(s) \right\} ds \\ &= \int_0^R \eta(s,k)f(s)ds , \end{aligned} \tag{10.21}$$

where we should note that we obtain from (10.3)

$$|G(r,s,k)f(s)|_X \leqq C(k)\ (1+s)^{1+\beta} |f(s)|_X , \tag{10.22}$$

and hence the dominated convergence theorem can be applied. Since f_R converges to f in $L_{2,\delta}(I.X)$ as $R\to\infty$ and $F(k)$ is a bounded linear operator from $L_{2,\delta}(I,X)$ into X, (10.18) is obtained by letting $n\to\infty$ in (10.21). If $f\in L_{2,\beta}(I,X)$ with $\beta > \frac{1}{2} + \delta$, then it follows from (10.12) that $|\eta(s,k)f(s)|_X$ is integrable over I. Therefore we obtain (10.19) from (10.18).

Q.E.D.

THEOREM 10.6. Let Assumption 5.1 be satisfied.

(i) Let $k\in\mathbf{R} - \{0\}$. Then $\eta^*(\cdot,k)x \in L_{2,-\delta}(I,X)$ for any $x \in X$ with the estimate

$$(10.23) \qquad \|\eta^*(\cdot,k)x\|_{-\delta} \leqq C|x|_X \qquad (x \in X),$$

where $C = C(k)$ is bounded when k moves in a compact set in $\mathbf{R} - \{0\}$.

(ii) We have

$$(10.24) \qquad (\eta^*(\cdot,k)x,f)_0 = (x,F(k)f)_X$$

for any triple $(k,x,f) \in (\mathbf{R} - \{0\})\times X \times L_2(I,X)$.

PROOF. (i) Let $g \in L_2(I,X)$ with compact support in $\bar{I}$. Then we obtain from Theorem 10.5

$$(10.25) \qquad (F(k)g,x)_X = (\int_I \eta(r,k)g(r)dr,\ x)_X = \int_I (g(r),\ \eta^*(r,k)x)_X dr .$$

Set in (10.25) $g(r) = \chi_R(r)(1+r)^{-2\delta}\eta^*(r,k)x$, $\chi_R(r)$ being the characteristic function of $(0,R)$. Then, noting that $F(k)$ is a bounded operator from $L_{2,\delta}(I,X)$ into X, we arrive at

$$\int_0^R (1+r)^{-2\delta} |\eta^*(r,k)x|_X^2 dr = (F(k)\{\chi_R(1+r)^{-2\delta}\eta^*(\cdot,k)x\},x)_X$$

(10.26) $$\leqq C(k) \, \|\chi_R(1+r)^{-2\delta}\eta^*(\cdot,k)x\|_\delta \, |x|_X$$
$$= C(k) \, \|\eta^*(\cdot,k)x\|_{-\delta,(0,R)} |x|_X \quad ,$$

which implies that

(10.27) $$\|\eta^*(\cdot,k)x\|_{-\delta,(0,R)} \leqq C(k) \, |x|_X \, .$$

Since $R > 0$ is arbitrary, (10.23) directly follows from (10.26). (ii) Let $f \in L_{2,\delta}(I,X)$. Set in (10.25) $g(r) = \chi_R(r)f(r)$ and let $R\to\infty$. Then, since $(f(r), \eta^*(r,k)x)_X$ is integrable over I by (10.23), we obtain (10.24), which completes the proof.

Q.E.D.

In order to show the continuity of $\eta(r,k)$ and $\eta^*(r,k)$ with respect to k we shall show

PROPOSITION 10.7. Let $x \in D$. Then we have

(10.28) $$\eta^*(r,k)x = \frac{1}{2ik}\, \xi(r)e^{i\mu(r\cdot,k)}x - h(r,k,x)$$
$$(r \in \overline{I},\ k \in \mathbb{R} - \{0\}),$$

where $\xi(r)$ is a real-valued smooth function on I such that $\xi(r) = 0$ $(r \leqq 1)$, $= 1$ $(r \geqq 2)$, $\mu(y,k)$ is as in (5.8) and $h(\cdot,k,x)$ is the radiative function for $\{L, -k, \mathcal{L}[\,f\,]\}$ with

(10.29) $$f(r) = \frac{1}{2ik}(L - k^2)(\xi e^{i\mu(r\cdot,k)}x) \quad .$$

PROOF. Let $w_0(r) = \xi(r)e^{i\mu(r\cdot,k)}x$ and $f_0=(L-k^2)w_0$ as in (8.30).

Then $f_0 \in L_{2,\delta}(I,X)$ and w_0 is the radiative function for

$\{L,k,\mathfrak{A}[f_0]\}$. Let $g \in C_0^\infty(I,X)$. Then, setting in (8.11) $f_1=f_0$, $f_2=g$

and making use of the fact that $F(k)f_0=x$ by Proposition 8.5, we have

(10.30)
$$(x,F(k)g)_x = \frac{1}{2ik} \{(w_0,g)_0 - (f_0,v)_0\} .$$

with the radiative function for $\{L,k,\mathfrak{A}[g]\}$.

By (ii) of Theorem 10.6 the left-hand side of (10.30) can be rewritten

as

(10.31)
$$(x,F(k)g)_x = (\eta*(\cdot,k)x, g)_0 .$$

As for the second term of the right-hand side we have, by exchanging

the order of integration,

$$(f_0,v)_0 = \int_0^\infty (f_0(r), \int_0^\infty G(r,s,k)g(s)ds)_x dr$$

(10.32)
$$= \int_0^\infty (\int_0^\infty G(s,r,-k)f_0(r)dr, g(s))_x ds$$

$$= (2ikh(\cdot),g)_0 ,$$

where we have used (10.2) repeatedly. (10.30) ~ (10.32) are combined

to give

(10.33)
$$(\eta*(\cdot,k)x,g)_0 = (\frac{1}{2ik} w_0 - h, g)_0 .$$

(10.29) follows from (10.33) because of the arbitrariness of $g \in C_0^\infty(I,X)$.

Q.E.D.

THEOREM 10.8. Let Assumption 5.1 be staisfied. Then $\eta(r,k)x$ and $\eta^*(r,k)x$ are strongly continuous X-valued functions on $\bar{I}\times(\mathbf{R}-\{0\})\times X$.

PROOF. Let us first show the continuity of $\eta^*(r,k)x$. To this end it is sufficient to show that $\eta^*(r_n,k_n)x_n$ tends to $\eta^*(r,k)x$ strongly in X, where $r_n\to r$ in $\bar{I}$, $k_n\to k$ in $\mathbf{R}-\{0\}$, $x_n\to x$ in X as $n\to\infty$. Let $\varepsilon>0$ given. Then, by the denseness of D and the estimate (10.12), there exists $x_0\in D$ and a positive integer n_0 such that

$$(10.34)\qquad \begin{cases} |\eta^*(r,k)x-\eta^*(r,k)x_0|<\varepsilon, \\ |\eta^*(r_n,k_n)x_n-\eta^*(r_n,k_n)x_0|<\varepsilon \end{cases}$$

for $n\geqq n_0$. Therefore we have only to show that

$$(10.35)\qquad s-\lim_{n\to\infty}\eta^*(r_n,k_n)x=\eta^*(r,k)x \qquad \text{in } X$$

for $x\in D$. In fact it follows from Proposition 10.7 that

$$(10.36)\qquad \begin{aligned} &\eta^*(r,k)x-\eta^*(r_n,k_n)x \\ &\qquad =\frac{1}{2ik}\{\xi(r)e^{i\mu(r\cdot,k)}-\xi(r_n)e^{i\mu(r_n\cdot,k_n)}\}x \\ &\qquad +\{h(r_n,k,x)-h(r,k,x)\} \\ &\qquad +\{h(r_n,k_n,x)-h(r_n,k,x)\}. \end{aligned}$$

Since $\mu(y,k)$ is continuous in (y,k) by (1) of Remark 5.9 and the definition of $\mu(y,k)$ (see (5.8)), the first term of the right-hand side of (10.36) tends to zero as $n\to\infty$. The second term of the right-hand side of (10.36) tends to zero because of the continuity of the radiative function $h(r,k,x)$. It follows from (1) of Remark 5.9 that $(L-k_n^2)(\xi e^{i\mu(r\cdot,k_n)}x)$ converges to $(L-k^2)(\xi e^{i\mu(r\cdot,k)}x)$ in $L_{2,\delta}(I,X)$, which implies that $h(\cdot,k_n,x)$ converges to $h(\cdot,k,x)$ in $L_{2,-\delta}(I,X) \cap H_0^{1,B}(I,X)_{loc}$. Since the convergence in $H_0^{1,B}(I,X)_{loc}$ means the uniform convergence in X on every finite interval $[0,R]$ on $\overline{I}$, we have

$$(10.37) \qquad s - \lim_{n\to\infty} \{h(r_n,k_n x) - h(r_n,k,x)\} = 0 \quad \text{in } X.$$

Thus (10.35) has be proved. The proof of the continuity of $\eta(r,k)x$ is much easier. In fact, the continuity follows from the facts that

$$(10.38) \qquad \eta(r,k)x = F(k)\ell[r,x] \qquad (r\in\overline{I},\ k\in\mathbf{R} - \{0\},\ x\in X),$$

and that as has been seen in the proof of Proposition 9.2, $\ell[r,x]$ is an F_δ-valued, continuous function on $\overline{I} \times X$ and that $F(k)$ is a $\mathbf{B}(F_\delta(I,X),X)$-valued, continuous function on $\mathbf{R} - \{0\}$.

Q.E.D.

§11. Expansion Theorem

Now we are in a position to show a spectral representation theorem or an expansion theorem for the Schrödinger operator with a long-range potential. To this end the self-adjoint realization of $T = -\Delta + \mathcal{Q}(y)$ should be defined and some of its spectral properties should be mentioned. Here let us note that these properties were never used in the previous sections (§1 - §10).

Let $\mathcal{Q}(y)$ be a real-valued, bounded and continuous function on $\mathbb{R}^N$. Then two symmetric operators h_0 and h in $L_2(\mathbb{R}^N)$ are defined by

$$
(11.1) \qquad \begin{cases} \mathcal{D}(h_0) = \mathcal{D}(h) = C_0^\infty(\mathbb{R}^N) , \\ h_0 f = T_0 f = -\Delta f \\ hf = Tf = -\Delta f + \mathcal{Q}(y)f . \end{cases}
$$

As is well-known, h_0 is essentially self-adjoint in $L_2(\mathbb{R}^N)$ with a unique self-adjoint extension H_0 defined by

$$
(11.2) \qquad \begin{cases} D(H_0) = H_2(\mathbb{R}^N) , \\ H_0 f = T_0 f = -\Delta f . \end{cases}
$$

where the differential operator T_0 should be considered in the distribution sense. Further, it is also well-known that we have

$$
(11.3) \qquad \sigma(H_0) = \sigma_{ac}(H_0) = [0,\infty) .
$$

Here $\sigma(H_0)$ is the spectrum of H_0 and $\sigma_{ac}(H_0)$ is the absolutely continuous spectrum.

THEOREM 11.1. Let $Q(y)$ be a real-valued, continuous function on $\mathbb{R}^N$ with $Q(y) \to 0$ as $|y| \to \infty$. Let h be as in (11.1). Then h is essentially self-adjoint in $L_2(\mathbb{R}^N)$ with its unique self-adjoint extension H. We have

$$(11.4) \qquad \begin{cases} \mathcal{D}(H) = \mathcal{D}(H_0) = H_2(\mathbb{R}^N) \\ Hf = Tf = -\Delta + Q(y)f \quad \text{(in the distribution sense).} \end{cases}$$

H is bounded below with the lower bound $k_0 \leqq 0$ and $\sigma_e(H)$, the essential spectrum of H, is equal to $[0,\infty)$. Therefore on $[k_0,0)$ the continuous spectrum of H is absent, and the negative eigenvalues, if they exist, are of finite multiplicity and are discrete in the sense that they form an isolated set having no limit point other than the origin 0. There exists no positive eigenvalue of H.

PROOF. Let $V = Q(y)\times$ be a multiplication operator $Q(y)$. Then V is a bounded operator on $L_2(\mathbb{R}^N)$ and h is rewritten as

$$(11.5) \qquad h = h_0 + V .$$

Since h_0 is essentially self-adjoint, h is also essentially self-adjoint with a unique self-adjoint extenstion $H = H_0 + V$ and the domain $\mathcal{D}(H)$ is equal to $\mathcal{D}(H_0)$ (see, e.g., Kato [1], p. 288, Theorem 4.4). V can be seen to be H_0-compact, i.e., if $\{f_n\} \subset D(H_0)$ and $\{H_0f_n\}$ are bounded sequence in $L_2(\mathbb{R}^N)$, then the sequence $\{Vf_n\}$ is relatively compact in $L_2(\mathbb{R}^N)$. In fact this can be easily shown by the interior estimate and the Rellich theorem. Therefore the essential spectrum $\sigma_e(H)$ of $H = H_0 + V$ is equal to the essential spectrum $\sigma_e(H_0)$ of H_0 (see Kato [1], p. 244, Theorem 5.35). Finally we shall show the non-existence of the positive eigenvalues of H. Suppose that $\lambda > 0$ is an eigenvalue of H and let $\varphi \in H_2(\mathbb{R}^N)$ be the eigenfunction associated with λ. Then $v = U\varphi$ is the radiative function for $\{L, \sqrt{\lambda}, 0\}$ and hence $v = 0$ by the uniqueness of the radiative function. This is a contradiction. Thus the non-existence of positive eigenvalues of H has been proved.

Q.E.D.

Set

$$
(11.6) \qquad \begin{cases} M_0 = UH_0U^* , \\ \\ M = UHU^* , \end{cases}
$$

where U is defined by (0.19). M_0 and M are self-adjoint operators in $L_2(I,X)$ with the same domain $UH_2(\mathbb{R}^N)$ and are unitarily equivalent to H_0 and H, respectively. Let $E(\cdot;M)$ be the spectral measure associated with M. We shall be mainly interested in the structure of the spectrum of M on $(0,\infty)$, because the spectrum on $(-\infty,0)$ is discrete.

<u>PROPOSITION 11.2</u>. Let Assumption 5.1 be satisfied. Let J be a compact interval in $(0,\infty)$ and let $f,g \in L_{2,\delta}(I,X)$. Then

$$(11.7)\qquad \begin{aligned}(E(J;M)f,g)_0 &= \int_{\sqrt{J}} \frac{2k^2}{\pi}(F(k)f,\ F(k)g)_X dk\\ &= \int_{\sqrt{J}} \frac{2k^2}{\pi}(F(-k)f,\ F(-k)g)_X dk\ ,\end{aligned}$$

where $\sqrt{J} = \{k > 0/k^2 \in J\}$ and $F(k)$ is given by Definition 8.2.

PROOF. We denote the resolvent of M by $R(z;M)$, i.e., $R(z;M) = (M - Z)^{-1}$. Let us note that $R(z;M)f = v(\cdot,\sqrt{z},\ell[f])$ for $z \in \mathbb{C} - \mathbb{R}$ and $f \in L_{2,}(I,X)$. Here $\sqrt{z}$ is the square root of z with $I_m\sqrt{z} \geqq 0$ and $v(\cdot,\sqrt{z},\ell[f])$ is the radiative function for $\{L,\sqrt{z},\ell[f]\}$. In fact this follows from the uniqueness of the radiative function and the fact that $R(z;M)f \in UH_2(\mathbb{R}^N)$. Moreover let us note that

$$(11.8)\qquad \lim_{b\downarrow 0} R(k^2 \pm ib;\ M)f = v(\cdot,\ \pm\ |k|,\ \ell[f]) \text{ in } L_{2,-\delta}(I,X)$$

$$(k \in \mathbb{R} - \{0\},\ f \in L_{2,\delta}(I,X))\ ,$$

which follows from the continuity of the radiative function with respect to $k \in \mathbb{C}^+$. Then from the well-known formula

$$(11.9)\qquad \begin{aligned}&(E(J;M)f,g)_0\\ &= \frac{1}{2\pi i}\lim_{b\downarrow 0}\int_J \{(R(A + ib;M)f,g)_0 - (f,R(a + ib;M)g)_0\}da\\ &= \frac{1}{2\pi i}\lim_{b\downarrow 0}\int_J \{(f,R(a - ib;M)g)_0 - (R(a - ib;M)f,g)_0\}da\end{aligned}$$

it follows that

$$(11.10)\qquad \begin{aligned}&(E(J;M)f,g)_0\\ &= \frac{1}{2\pi i}\int_J \{(v(\cdot,\sqrt{a},\ell[f]),g)_0 - (f,v(\cdot,\sqrt{a},\ell[g]))_0\}da\\ &= \frac{1}{2\pi i}\int_J \{(f,v(\cdot,-\sqrt{a},\ell[g]))_0 - (v(\cdot,-\sqrt{a},\ell[f]),g)_0\}da\ .\end{aligned}$$

By the use of (8.14) in Theorem 8.4 (11.7) follows from (11.10).

Q.E.D.

Let us set

(11.11) $$F_{\pm}(k) = \pm\sqrt{\frac{2}{\pi}}\, ikF(\pm k) \qquad (k > 0)\ .$$

COROLLARY 11.3. (i) $E((0,\infty);M)M$ is an absolutely continuous operator, i.e.,

$$\sigma_{ac}(M) \supset (0,\infty)\ .$$

(ii) $F_{\pm}(\cdot)f \in L_2(0,\infty),X,dk)$ for $f \in L_{2,\delta}(I,X)$ with the estimate

(11.13) $$\int_0^{\infty} |F_{\pm}(k)f|_X^2\, dk = \| E((0,\infty);M)f\|_0^2 \leq \| f\|_0^2\ .$$

PROOF. Obvious from (11.7) and the denseness of $L_{2,\delta}(I,X)$ in $L_{2,}(I,X)$.

Q.E.D.

DEFINITION 11.4. The *generalized Fourier transforms* $F_{\pm}$ from $L_2(I,X) = L_2(I,X,dr)$ into $L_2((0,\infty),X,dk)$ by

(11.14) $$(F_{\pm}f)(k) = \underset{R\to\infty}{\text{l.i.m.}}\ F_{\pm}(k)f_R \quad \text{in } L_2((0,\infty),X,dk)$$

for $f \in L_2(I,X)$, where $f_R(r) = \chi_R(r)f(r)$ and $\chi_R(r)$ is the characteristic function of the interval $(0,R)$.

It follows from Corollary 11.3 that $F_{\pm}$ are bounded linear operators from $L_2(I,X)$ into $L_2((0,\infty),X,dk)$. $F_{\pm}^{*}$ denote the adjoint operators from $L_2((0,\infty),X,dk)$ into $L_2(I,X)$. By the use of the relation between $F(k)$ and $\eta(r,k)$. Let us set

(11.15) $$\begin{cases} \eta_{\pm}(r,k) = \pm\sqrt{\frac{2}{\pi}}\, ik\eta\,(r,\pm k) \\ \eta_{\pm}^{*}(r,k) = \mp\sqrt{\frac{2}{\pi}}\, ik\eta^{*}(r,\pm k) \end{cases}$$

for $r \in \overline{I}$, $k > 0$.

PROPOSITION 11.5. Let Assumption 5.1, (5.4) and (5.5) be satisfied. Let $F_{\pm}$ be as above.

(i) Then $F_{\pm}$ and $F_{\pm}^{*}$ are represented as follows:

$$(11.16)\quad (F_{\pm}f)(k) = \underset{R\to\infty}{\text{l.i.m.}} \int_0^R \eta_{\pm}(r,k)f(r)dr \quad \text{in } L_2((0,\infty),X,dk)$$

$$(11.17)\quad (F_{\pm}^{*}F)(k) = \underset{R\to\infty}{\text{l.i.m.}} \int_{R^{-1}}^R \eta_{\pm}^{*}(r,k)F(k)dk \quad \text{in } L_2(I,X),$$

where $f \in L_2(I,X)$ and $F \in L_2((0,\infty),X,dk)$.

(ii) Let $f \in L_{2,\delta}(I,X)$. Then for each $k > 0$

$$(11.18)\qquad (F_{\pm}f)(k) = s - \lim_{R\to\infty} \int_0^R \eta_{\pm}(r,k)f(r)dr \quad \text{in } X.$$

In particular, if $f \in L_{2,\beta}(I,X)$ with $\beta > \frac{1}{2} + \delta$, the integral of the right-hand side of (11.18) is absolutely convergent.

PROOF. Let $f_R(r) = \chi_R(r)f(r)$ with the characteristic function $\chi_R(r)$ of $(0,R)$. Then it follows from (10.19) in Theorem 10.5 that

$$(11.19)\quad \begin{aligned} F_{\pm}(k)f_R &= \pm\sqrt{\tfrac{2}{\pi}}\, ikF(\pm ik)f = \pm\sqrt{\tfrac{2}{\pi}}\, ik \int_I \eta(r,\pm k)f_R(r)dr \\ &= \int_0^R \eta_{\pm}(r,k)f(r)dr, \end{aligned}$$

which, together with the definition of $F_{\pm}f$ ((11.14)), proves (11.16). In order to prove (11.17) it suffices to show

$$(11.20)\qquad (F_{\pm}^{*}F)(r) = \int_0^{\infty} \eta_{\pm}^{*}(r,k)F(k)dk$$

for $F \in L_2((0,\infty),X,dk)$ with compact support in $(0,\infty)$, because $F_{\pm}^{*}$ are bounded linear operators from $L_2((0,\infty),X,dk)$ into $L_2(I,X)$. Denoting the inner product of $L_2((0,\infty), X,dk)$ by $(\ ,\ )_0$ again, we have for any $f \in C_0^\infty(I,X)$.

$$
\begin{aligned}
(F_\pm F, f)_0 &= (F, \pm f)_0 \\
&= \int_0^\infty (F(k)\ , \int_0^\infty \eta_\pm(r,k) f(r)dr)_X dk \\
&= \int_0^\infty \Big(\int_0^\infty \eta_\pm^*(r,k)F(k)dk\ ,\ f(r) \Big)_X dr \\
&= \Big(\int_0^\infty \eta_\pm^*(\cdot,k)F(k)dk\ ,\ f \Big)_0 ,
\end{aligned}
$$

which implies (11.17). Here it should be noted that the integrals $\int_0^\infty \eta_\pm^*(r,k)F(k)dk$ and $\int_0^\infty \eta_\pm(r,k)f(r)dr$ are continuous in r and in k, respectively, because of Theorem 10.8. (ii) directly follows from Theorem 10.5 and Definition 11.4.

Q.E.D.

Thus we arrive at

<u>THEOREM 11.6</u> (Spectral Representation Theorem). Let Assumption 5.1 be satisfied. Let B be an arbitrary Borel set in $(0,\infty)$. Then

(11.21) $$E(B;M) = F_\pm^* \chi_{\sqrt{B}}\ F_\pm ,$$

where $\chi_{\sqrt{B}}$ is the characteristic function of $\sqrt{B} = \{k > 0 / k^2 \in B\}$. In particular we have

(11.22) $$E(0,\infty);M) = F_\pm^* F_\pm .$$

<u>PROOF</u>. It suffices to show (11.21) when B is a compact interval J in $(0,\infty)$. From (11.7) it follows that

$$
\begin{aligned}
(E(J;M)f,g)_0 &= \int_{\sqrt{J}} ((F_\pm f)(k)\ ,\ (F_\pm g)(k))_X dk \\
&= (\chi_{\sqrt{J}}(F_\pm f)(\cdot)\ ,\ (F_\pm g)(\cdot))_0 \qquad (11.23)\\
&= (F_\pm^*\ \chi_{\sqrt{J}} F_\pm f, g)_0
\end{aligned}
$$

for $f,g \in L_{2,\delta}(I,X)$, which implies that

(11.24) $$E(J;M)f = F_\pm^{*}\chi_{\sqrt{J}}F_\pm f \qquad (f \in L_{2,\delta}(I,X)$$

Since $E(J;M)$ and $F_\pm \chi_{\sqrt{J}} F_\pm$ are bounded linear operators on $L_2(I,X)$ and $L_{2,\delta}(I,X)$ is dense in $L_2(I,X)$, (11.21) follows from (11.24).

Q.E.D.

In order to discuss the orthogonality of the generalized Fourier transforms $F_\pm$, some properties of $F_\pm$ will be shown.

PROPOSITION 11.7. Let $F_\pm$ be the generalized Fourier transforms associated with M.

(i) Then

(11.25) $$F_\pm E(B;M) = \chi_{\sqrt{B}} F_\pm ,$$

(11.26) $$F_\pm E((0,\infty);M) = F_\pm$$

(11.27) $$E(B;M)F_\pm^* = F_\pm^*\chi_{\sqrt{B}} ,$$

(11.28) $$E((0,\infty);M)F_\pm^* = F_\pm^*,$$

where B and $\chi_{\sqrt{B}}$ are as in Theorem 11.6.

(ii) $F_\pm L_2(I,X)$ are closed linear subspace in $L_2((0,\infty),X,dk)$.

(iii) The following three condition (a), (b), (c) about F_+ (F_-) are equivalent:

(a) F_+ (F_-) maps $E((0,\infty);M)L_2(I,X)$ onto $L_2((0,\infty),X,dk)$.

(b) $F_+F_+^* = 1$ $(F_-F_-^* = 1)$, where 1 means the identity operator.

(c) The null space of F_+ (F_-) consists only of 0.

PROOF. (i) Let us show (11.25). Let $f \in L_2(I,X)$. Then

$$K = \| F_\pm E(B;M)f - \chi_{\sqrt{B}} F_\pm f \|_0^2$$

$$(11.29) = \|F_\pm E(B;M)f\|_0^2 + \|\chi_{\sqrt{B}} F_\pm f\|_0^2 - 2\mathrm{Re}\,(F_\pm E(B;M)f, \chi_{\sqrt{B}} F_\pm f)$$

$$= J_1 + J_2 - 2\mathrm{Re}J_3 .$$

From (11.21) and the relations

$$(11.30) \qquad E(B;M)^2 = E((0,\infty);M)E(B;M) = E(B;M)$$

we can easily see that $J_1 = J_2 = J_3$, and hence $K = 0$, which completes the proof of (11.25). (11.26)-(11.28) immediately follows from (11.25). (ii) Assume that the sequence $\{F_+f_n\}$, $f_n \in L_2(I,X)$, be a Cauchy sequence in $L_2((0,\infty),X,dk)$. By taking account of (11.26) f_n may be assumed to belong to $E((0,\infty);M)L_2(I,X)$. Then $\{F_+^*F_+f_n\}$ is a Cauchy sequence, since F_+^* is a bounded operator. Thus, by noting that $F_+^*F_+f_n = E((0,\infty);M)f_n = f_n$, the sequence $\{f_n\}$ itself is a Cauchy sequence in $L_2(I,X)$ with the limit $f \in L_2(I,X)$. Therefore the sequence $\{F_+f_n\}$ has the limit F_+f , which means the closedness of $F_+L_2(I,X)$ in $L_2((0,\infty),X,dk)$. The closedness of $F_-L_2(I,X)$ can be proved quite in the same way. (iii) By the use of (i) and (ii) the equivalence of (a), (b), (c) can be shown in an elementary and usual way, and hence the proof will be left to the readers.

Q.E.D.

When the generalized Fourier transform F_+ or F_- satisfies one of the conditions (a), (b), (c) in Proposition 11.7, it is called *orthogonal*. The notion of the orthogonality is important in scattering theory.

THEOREM 11.8 (orthogonality of $F_\pm$). Let Assumptions 5.1, (5.4), and (5.5) be satisfied. Then the generalized Fourier transforms are orthogonal.

PROOF. We shall show that $F_{\pm}$ satisfy (c) in Proposition 11.7. Suppose that $F_{+}^{*}F = 0$ with some $F \in L_2((0,\infty),X,dk)$. We have only to show that $F = 0$. Let $B = (a^2,b^2)$ with $0 < a < b < \infty$. Then, using (11.27), we obtain from $E(B;M)F_{+}^{*}F = 0$

$$F^{*}\chi_{\sqrt{B}} F = 0 , \tag{11.31}$$

which, together with (11.17), gives

$$\int_a^b \eta_+^{*}(r,k)F(k)dk = 0 . \tag{11.32}$$

Since a and b can be taken arbitrarily, it follows that there exists a null set e in $(0,\infty)$ such that

$$\eta_+^{*}(r,k)F(k) = 0 \quad \text{or} \quad \eta^{*}(r,k)F(k) = 0 \qquad (r \notin \overline{I},\ k \in e) , \tag{11.33}$$

where we have made use of the continuity of $\eta_+^{*}(r,k)F(k)$ in $r \in I$ (Theorem 10.8). Let $x \in D$ and take $f_0(r)$ as in (8.30). Then, noting that $F(k)f_0 = x$ by Proposition 8.5 and using (ii) of Theorem 10.6, we have

$$\begin{aligned} (F(k),x)_X &= (F(k),\ F(k)f_0)_X \\ &= (\eta^{*}(\cdot,k)F(k),\ f_0)_0 = 0 \qquad (k \notin e,\ x \in D) \end{aligned} \tag{11.34}$$

whence follows that $F(k) = 0$ for almost all $k \in (0,\infty)$ i.e., $F = 0$ in $L_2((0,\infty),X,dk)$. The orthogonality of F_- can be proved quite in the same way.

Q.E.D.

Finally we shall translate the expansion theorem, Theorem 11.6 for the operator M into the case of the Schrödinger operator H . Let us define the generalized Fourier transforms $\tilde{F}_{\pm}$ associated with H by

(11.35) $$\tilde{F}_{\underline{+}} = U_k^{-1} F_{\underline{+}} U ,$$

where $U_k = k^{\frac{N-1}{2}} x$ is a unitary operator from $L_2(\mathbb{R}^N, d\xi)$ onto $L_2((0,\infty),X,dk$. $\tilde{F}_{\underline{+}}$ are bounded linear operators from $L_2(\mathbb{R}^N,dy)$ into $L_2(\mathbb{R}^N,d\xi)$. If the bounded operators $\tilde{\eta}_{\pm}(r,k)$ and $\tilde{\eta}_{\pm}^*(r,k)$, $r \in I$, $k > 0$, on X are defined by

(11.36)
$$\begin{cases} \tilde{\eta}_{\pm}(r,k) = r^{-(N-1)/2} k^{-(N-1)/2} \eta_{\underline{+}}(r,k) , \\ \tilde{\eta}_{\pm}^*(r,k) = r^{-(N-1)/2} k^{-(N-1)/2} \eta_{\underline{+}}^*(r,k) , \end{cases}$$

then we have

(11.37)
$$\begin{aligned} (\tilde{F}_{\underline{+}}\Phi)(\xi) &= \underset{R\to\infty}{\mathrm{l.i.m.}} \int_0^R (\tilde{\eta}_{\underline{+}}(r,k)\Phi(r\cdot))(\omega')dr \qquad \text{in } L_2(\mathbb{R}^N,d\xi) \\ (\tilde{F}_{\underline{+}}^*\Psi)(y) &= \underset{R\to\infty}{\mathrm{l.i.m.}} \int_{R^{-1}}^R (\tilde{\eta}_{\underline{+}}^*(r,k)\Psi(k\cdot))(\omega)dk \qquad \text{in } L_2(\mathbb{R}^N,dy) \end{aligned}$$

where $\tilde{F}_{\underline{+}}^*$ are the adjoint of $\tilde{F}_{\underline{+}}$ and $y = r\omega$, $\xi = k\omega'$. Let $E(\cdot;H)$ be the spectral measure associated with H . Then we have from Theorems 11.6 and 11.8.

<u>THEOREM 11.9</u>. Let Assumption 5.1, (5.4) and (5.5) be satisfied. Let B be an arbitrary Borel set in $(0,\infty)$. Then the generalized Fourier transforms $\tilde{F}_{\underline{+}}$, defined as above, map $E((0,\infty);H)L_2(\mathbb{R}^N,dy)$ onto $L_2(\mathbb{R}^N,d\xi)$ and

(11.38) $$E(B;H) = \tilde{F}_{\underline{+}}^* \tilde{\chi}_{\sqrt{B}} \tilde{F}_{\underline{+}} ,$$

where $\tilde{\chi}_{\sqrt{B}}$ is the characteristic function of the set $\{\xi \in \mathbb{R}^N/$
$\{\xi \in \mathbb{R}^N/\ |\xi|^2 \in B\}$.

§12. The General Short-Range Case

In this section we shall consider the case that the short-range potential $\mathcal{Q}_1$ is a general short-range potential, i.e., $\mathcal{Q}_1(y) = o(|y|^{-1-\varepsilon})$. Throughout this section we shall assume

ASSUMPTION 12.1. The potential $\mathcal{Q}(y) = \mathcal{Q}_0(y) + \mathcal{Q}_1(y)$ satisfies $(\mathcal{Q})$, $(\mathcal{Q}_1)$ in Assumption 2.1. Further, $\mathcal{Q}_0(y)$ is assumed to satisfy $(\tilde{\mathcal{Q}}_0)$ in Theorem 5.1.

Then all the results of Chapter I and §9 of Chapter III are valid in the present case. Therefore the limiting absorption principle holds good and the Green kernel $G(r,s,k)$ is well-defined. In order to make use of the results of Chapter II and §10 - §11 of Chapter III, we shall approximate $\mathcal{Q}_1(y)$ by a sequence $\{\mathcal{Q}_{1n}(y)\}$ which satisfies $|\mathcal{Q}_{1n}(y)| = \mathcal{O}(|y|^{-\varepsilon_1})$ $(n=1,2,\ldots)$ with a constant $\varepsilon_1 > \max(2 - \varepsilon, \frac{3}{2})$. To this end let us define $\mathcal{Q}_{1n}(y)$ by

$$\mathcal{Q}_{1n}(y) = \varphi(|y| - n)\mathcal{Q}_1(y) \qquad (n=1,2,\ldots), \tag{12.1}$$

where $\varphi(t)$ is a real-valued continuous function on $\mathbb{R}$ such that $\varphi(t) = 1$ $(t \leqq 0)$, $= 0$ $(t \geqq 1)$, and let us set

$$\begin{cases} T_n = -\Delta + \mathcal{Q}_0(y) + \mathcal{Q}_1(y), \\ L_n = -\dfrac{d^2}{dr^2} + B(r) + C_0(r) + C_{1n}(r) \qquad (C_{1n}(r) = \mathcal{Q}_{1n}(r\omega)\times). \end{cases} \tag{12.2}$$

Since the support of $\mathcal{Q}_{1n}(y)$ is compact, all the result in the preceding sections can be applied to L_n. According to Definition 8.3,

$F_n(k)$, $R \in \mathbb{R} - \{0\}$, is well-defined as a bounded linear operator from $F_\delta(I,X)$ into X . The eigenoperator $\eta(r,k)(r \in \bar{I}$, $k \in \mathbb{R} - \{0\})$ is also well-defined. Here we should note that the stationary modifier $\lambda(y,k)$ and its kernel $Z(y,k)$ are common for all L_n, $n=1,2,\ldots$, because the long-range potential $Q_0(y)$ is independent of n .

We shall first show that an operator $F(k)$ associated with L is well-defined as the strong limit of $F_n(k)$ and that $F(k)$ satisfies all the properties obtained in §8.

PROPOSITION 12.2. Let Assumption 12.1 be satisfied. Let $F_n(k)(k \in \mathbb{R}-\{0\}, n=1,2,\ldots)$ be the bounded linear operator from $F_\delta(I,X)$ into X defined as above.

(i) Then the operator norm $\|F_n(k)\|$ is uniformly bounded when $n=1,2,\ldots$ and k moves in a compact set in $\mathbb{R}-\{0\}$. For each $k \in \mathbb{R}-\{0\}$, there exists a bounded linear operator $F(k)$ from $F_\delta(I,X)$ into X such that

$$(12.3) \qquad F(k)\ell = s - \lim_{n\to\infty} F_n(k)\ell \qquad (\ell \in F_\delta(I,X))$$

in X . $\|F(k)\|$ is bounded when k moves in a compact set in $\mathbb{R}-\{0\}$.

(ii) $F(k)$ satisfies (8.12) for any $\ell_1,\ell_2 \in F_\delta(I,X)$, i.e., we have

$$(12.4) \qquad (F(k)\ell_1,F(k)\ell_2)_X = \frac{1}{2ik}\{\overline{<\ell_2,v_1>} - <\ell_1,v_2>\} ,$$

$v_j(j = 1,2)$ being the radiative function for $\{L,k,v_j\}$. (8.13)-(8.15) are also satisfied by $F(k)$.

(iii) Let $x \in D$ and let $f_0(r)$ be as in (8.30), i.e., $f_0(r) = (L - k^2)(\xi e^{i\mu(r\cdot,k)}x)$ with $\xi(r)$ defined by (5.36) and $\mu(y,k)$ defined by (5.8) . Then

$$(12.5) \qquad \mathcal{F}(k)f_0 = x ,$$

where we set $\mathcal{F}(k)\ell[f_0] = \mathcal{F}(k)f_0$.

(iv) As an operator from $L_{2,\delta}(I,X)$ into X , $\mathcal{F}(k)$ is a $\mathbb{B}(L_{2,\delta}(I,X),X)$-valued continuous function on $\mathbb{R} - \{0\}$. Further, for each $k \in \mathbb{R} - \{0\}$ $\mathcal{F}(k)$ is a compact operator from $L_{2,\delta}(I,X)$ into X .

(v) Let $f \in L_{2,\delta}(I,X)$ and let $k \in \mathbb{R} - \{0\}$. Then

$$(12.6) \qquad \mathcal{F}(k)f = s - \lim_{n\to\infty} e^{-i\mu(r_n\cdot,k)}v(r_n) \quad \text{in } X ,$$

where v is the radiative function for $\{L,k,\ell[f]\}$ and $\{r_n\}$ is a sequence such that $r_n \uparrow \infty$ and $v'(r_n) - ikv(r_n) \to 0$ in X as $n \to \infty$.

PROOF. Proceeding as in the proof of Theorem 8.4 we obtain

$$(12.7) \qquad \begin{aligned} &(\mathcal{F}_n(k)\ell, \mathcal{F}_p(k)\ell)_X \\ &= \frac{1}{2ik}\{<\overline{\ell,v_n}> - <\ell,v_p> + (\{c_{1n} - c_{1p}\}v_n,v_p)_0\} \end{aligned}$$

for $\ell \in F_\delta(I,X)$ and $n,p=1,2,\ldots,$ where $v_n(v_p)$ is the radiative function for $\{L_n,k,\ell\}$ ($\{L_p,k,\ell\}$. Set $p = n$ in (12.7) and make use of Theorem 4.5. Then we have

$$(12.8) \qquad |\mathcal{F}_n(k)\ell|_X^2 \leqq \frac{1}{|R|} |||\ell|||_\delta \| v_n \|_{B,-\delta} \leqq C^2|||\ell|||_\delta^2$$

with $C = C(k)$ which is bounded when k moves in a compact set in $\mathbb{R} - \{0\}$, where $\| \ \|_{B,-\delta}$ is the norm of $H^{1,B}_{0,-\delta}(I,X)$. On the other hand we obtain from (12.7)

$$(12.9) \qquad \begin{aligned} |\mathcal{F}_n(k)\ell - \mathcal{F}_p(k)\ell|_X^2 &= |\mathcal{F}_n(k)\ell|_X^2 + |\mathcal{F}_p(k)\ell|_X^2 - 2\mathrm{Re}(\mathcal{F}_n(k)\ell,\mathcal{F}_p(k)\ell) \\ &= -\frac{1}{k} I_m(\{C_{1n} - C_{1p}\}v_n,v_p)_0 \to 0 \end{aligned}$$

as $n,p \to \infty$, because by Theorem 4.5 both v_n and v_p converge to the radiative function v for $\{L,k,\ell\}$ in $L_{2,-\delta}(I,X)$ and $C_{1n}(r) - C_{1p}(r)$ satisfies

$$\|C_{1n}(r) - C_{1p}(r)\| = |\varphi(r-n) - \varphi(r-p)|\ \|C_1(r)\| < c_0(1+r)^{-\varepsilon_1}\ (r \in \bar{I})$$
(12.9)
$$\|C_{1n}(r) - C_{1p}(r)\| \to 0 \qquad (n,p \to \infty)\ .$$

Therefore the strong limit $F(k)\ell = \underset{n\to\infty}{\text{s-lim}}\, F_n(k)$ exists in X and we have the estimate $\|F(k)\| \leqq C$, where C is as in (12.8), whence (i) follows. (12.4) immediately follows by letting $n \to \infty$ in the relation

$$(12.11)\quad (F_n(k)\ell_1, F_n(k)\ell_2)_X = \frac{1}{2ik}\{\overline{<\ell_2, v_{n1}>} - <\ell_1, v_{n2}>\}\ ,$$

$v_{nj}, j=1,2,$ being the radiative function for $\{L_n,k,\ell_j\}$ and Theorem 4.5 being made use of. Thus (ii) has been proved. Let us show (iii). Set

$$(12.12)\qquad f_{0n}(r) = (L_n - k^2)(\xi e^{i\mu(r\cdot,k)}x) \qquad (n=1,2,\ldots)\ .$$

Then it follows from Proposition 8.5 that

$$(12.13)\qquad F_n(k)f_{0n}(r) = x \qquad (n=1,2,\ldots)\ .$$

Further, it is easy to see from (5.20) in Lemma 5.5 that

$$(12.14)\qquad f_{0n} \to f_0 \quad (n \to \infty) \quad \text{in } L_{2,\delta}(I,X)\ .$$

(12.5) is obtained by letting $n \to \infty$ in (12.13) and using (i) and (12.14) Re-examining the proof of Theorems 8.6 and 8.7, we can see that (iv) and (v) are obtained by proceeding as in the proof of Theorems 8.6 and 8.7, respectively. This completes the proof.

Q.E.D.

Let us now define the eigenoperator $\eta(r,k)$ associated with L by

$$(12.15)\quad \eta(r,k)x = F(k)\ell[r,x] \qquad (r \in \bar{I}\ ,\ k \in \mathbb{R} - \{0\}\ ,\ x \in X)\ ,$$

where $\ell[r,x]$ is as in (9.2), i.e., $\langle\ell[r,x],\phi\rangle = (x,\phi(r))_X$ for $\phi \in H^{1,B}_0(I,X)$. At the same time, as will be shown in the following proposition, $\eta(r,k)$ can be defined as the strong limit of the eigenoperator $\eta_n(r,k)$ associated with L_n .

<u>PROPOSITION 12.3.</u> Let Assumption 12.1 be satisfied. Let $\eta(r,k)$ be the eigenoperator associated with L .

(i) Then for each $(r,k) \in \bar{I} \times (\mathbb{R} - \{0\})$ $\eta(r,k)$ is a bounded

$$\|\eta(r,k)\| \leqq C(1 + r)^{\delta} \qquad (r \in \bar{I}) \tag{12.16}$$

with $C = C(k)$ which is bounded when k moves in a compact set in $\mathbb{R} - \{0\}$. $\eta(r,k)x$ is an X-valued, continuous function on $\bar{I} \times (\mathbb{R} - \{0\}) \times X$.

(ii) We have

$$2ik(\eta(s,k)x,\eta(r,k)x')_X = (\{G(r,s,k) - G(r,s,-k)\}x,x')_X \tag{12.17}$$

for any $x,x' \in X$, $k \in \mathbb{R} - \{0\}$, and $r,s,\in \bar{I}$.

(iii) The operator norm $\|\eta_n(r,k)\|$ is estimated as

$$\|\eta_n(r,k)\| \leqq C(1 + r)^{\delta} \qquad (r \in \bar{I}, n=1,2,\ldots) \tag{12.18}$$

with $C = C(k)$ which is bounded when R moves in a compact set in $\mathbb{R} - \{0\}$. Further, we have

$$\eta(r,k)x = s - \lim_{n\to\infty} \eta_n(r,k)x \quad \text{in} \quad X \tag{12.19}$$

for any triple $(r,k,x) \in \bar{I} \times (\mathbb{R} - \{0\}) \times X$.

(iv) Let $f \in L_{2,\delta}(I,X)$ and let $k \in \mathbb{R} - \{0\}$. Then

$$(12.20) \qquad F(k)f = s - \lim_{R\to\infty}\int_0^R \eta(r,k)f(r)dr \quad \text{in } X .$$

PROOF. (i) and (ii) can be easily obtained by proceeding as in the proof of Theorems 10.4 and 10.8. Since we have

$$(12.21) \qquad \eta_n(r,k)x = F_n(k)\ell[s,k] ,$$

the uniform boundedness of $\|\eta_n(r,k)\|$ and the strong convergence of $\eta_n(r,k)x$ to $\eta(r,k)x$ follows from Proposition 12.2. Let us show (iv). It is sufficient to show

$$(12.22) \qquad F(k)f = \int_I \eta(r,k)f(r)dr$$

for $f \in L_2(I,X)$ with compact support in $\bar{I}$. Applying Theorem 10.5 to L_n, we obtain

$$(12.23) \qquad F_n(k) = \int_I \eta_n(r,k)f(r)dr ,$$

whence (12.22) follows by letting $n \to \infty$.

Q.E.D.

Let $\eta^*(r,k)$ be the adjoint of $\eta(r,k)$. Almost all the results with respect to $\eta^*(r,k)$ obtained in §10 are also valid for our $\eta^*(r,k)$.

PROPOSITION 12.4. Let Assumption 12.1 be satisfied. Let $\eta^*(r,k)$ be as above and let $\eta_n^*(r,k)$ be the adjoint of the eigenoperator $\eta_n(r,k)$ associated with L_n.

(i) Let $R \in \mathbb{R}-\{0\}$. Then $\eta^*(\cdot,k)x \in L_{2,-\delta}(I,X)$ for any $x \in X$ with the estimate

$$(12.24) \qquad \|\eta^*(\cdot,k)x\|_{-\delta} \leqq C|x|_X \quad (x \in X) ,$$

where $C = C(k)$ is bounded when k moves in a compact set in $\mathbb{R}-\{0\}$.
We have

$$(\eta^*(\cdot,k)x,f)_0 = (x,F(k)f)_X$$

(12.25)

$$(k \in \mathbb{R}-\{0\},\ x \in X,\ f \in L_{2,\delta}(I,X)) .$$

(ii) Let $x \in D$. Then

(12.26) $$\eta^*(r,k)x = \frac{1}{2ik}\,\xi(r)e^{i\mu(r\cdot,k)}x - h(r,k,x) .$$

where $\xi(r)$ is as in Proposition 10.7 and $h(\cdot,k,x)$ is the radiative function for $\{L,-k,\ell[f_0]\}$ with

(12.27) $$f_0(r) = \frac{1}{2ik}\,(L - k^2)(\xi e^{i\mu(r\cdot,k)}x) .$$

(iii) There exists $C = C(k)$, which is bounded when k moves in a compact set in $\mathbb{R}-\{0\}$, such that

(12.28) $$\|\eta_n^*(\cdot,k)x\|_{-\delta} \leqq C|x|_X \qquad (x \in X) .$$

$\{\eta_n^*(\cdot,k)x\}$ converges to $\eta^*(\cdot,k)x$ in $L_{2,-\delta}(I,X) \cap H^{1,B}_0(I,X)_{loc}$ for each $x \in X$.

(iv) $\eta^*(\cdot,k)x$ $H^{1,B}_0(I,X)_{loc} \cap D(I)$ and satisfies the equation

(12.29) $$(L - k^2)v(r) = 0 \qquad (r \in \bar{I}, k \in \mathbb{R}-\{0\}, x \in X) .$$

$\eta^*(r,k)x$ is an x-valued, continuous function on $\bar{I} \times (\mathbb{R}-\{0\}) \times X$.

PROOF. (i) and (ii) can be obtained by proceeding as in the proof of Theorem 10.6 and Proposition 10.7, respectively. Let us turn into the proof of (iii). The uniform boundedness of $\|\eta_n(\cdot,k)x\|_{-\delta}$ follows from the relation

(12.30) $$(\eta_n^*(\cdot,k)x, f)_0 = (x, F_n(k)f)_X$$

$$(f \in L_{2,\delta}(I,X),\ n = 1,2,\ldots)$$

and the uniform boundedness of $\|F_n(k)\|$. Now that (10.28) has been established, in order to show the convergence of $\{\eta_n^*(\cdot,k)x\}$ to $\eta^*(\cdot,k)x$ in $L_{2,-\delta}(I,X)$ it is sufficient to show for $x \in D$

(12.31) $$\eta_n^*(\cdot,k)x \to \eta^*(\cdot,k)x \quad (k \in \mathbb{R}-\{0\})$$

in $L_{2,-\delta}(I,X)$. Applying Proposition 10.7 to L_n, we have

(12.32) $$\eta_n^*(r,k)x = \frac{1}{2ik}\,\xi(r)e^{i\mu(r\cdot,k)}x - h_n(r,k,x)$$

for $x \in D$, $(r,k) \in \bar{I} \times \mathbb{R}-\{0\}$, where $h_n(\cdot,k,x)$ is the radiative function for $\{L_n, -k, \ell[f_n]\}$ and $f_{0,n} = (L_n - k^2)(\xi e^{i\mu(r\cdot,k)}x)$. Since $\{f_{0,n}\}$ can be easily seen to converge to $f_0 = (L - k^2)(\xi e^{i\mu(r\cdot,k)}x)$ in $L_{2,\delta}(I,X)$, Theorem 4.5 is applied to show that $\{h_n(\cdot,k,x)\}$ converges to $h(\cdot,k,x)$ in $L_{2,-\delta}(I,X)$, which implies (10.31). Noting that $\eta_n^*(r,k)x$ satisfies the equation $(L_n - k^2)v = 0$, we make use of the interior estimate and the convergence of $\{\eta_n^*(\cdot,k)x\}$ in $L_{2,-\delta}(I,X)$ to show the convergence of $\{\eta_n^*(\cdot,k)x\}$ in $H_0^{1,B}(I,X)_{loc}$, which completes the proof of (iii). We have

(12.33) $$(\eta^*(\cdot,k)x,\ (L - k^2)\phi)_0 = 0$$

$$(k \in \mathbb{R}-\{0\},\ x \in X,\ \phi \in C_0^\infty(I,X))$$

by letting $n \to \infty$ in the relation

(12.34) $$(\eta_n^*(\cdot,k)x,\ (L_n - k^2)\phi)_0 = 0$$

$$(n=1,2,\ldots,k \in \mathbb{R}-\{0\}, x \in X, \phi \in C_0^\infty(I,X))\ .$$

The first half of (iv) follows from (10.31) and Proposition 1.3. The continuity of $\eta^*(r,k)x$ can be shown in quite the same way as in the proof of Theorem 10.8, which completes the proof.

Q.E.D.

Now the Propositions 12.2, 12.3, 12.4 have been established, we can show the expansion theorem (Theorem 11.6) with the orthogonality of the generalized Fourier transforms $F_\pm$ (Theorem 11.8) in quite the same way as in the proof of Theorems 11.6 and 11.8. Therefore we shall omit the proof of the following

THEOREM 12.5. Let Assumption 12.1 be satisfied. Then the generalized Fourier transforms $F_\pm$ is well-defined by Definition 11.4. $F_\pm$ and the adjoint $F_\pm^*$ can be represented as (11.16) and (11.17), respectively. $F_\pm$ are orthogonal and (11.21) holds good.

Concluding Remarks

1° For the proof of the unique continuation theorem (Proposition 1.4) we referred to the unique continuation theorem for partial differential equation. On the other hand Jäger [3] gives the unique continuation theorem for an abstract ordinary differential operator with operator-valued coefficients which can be referred to in our case.

2° Our proof of the limiting absorption principle was along the line of Saitō [3]. In Ikebe-Saitō [1] and Lavine [1] the Schrödinger operator was directly treated and the limiting absorption principle for the Schrödinger operator with a long-range potential was proved.

3° In §5 we introduced the kernel $Z(y,k)$ of the stationary modifier $\lambda(y,k)$ as a solution of the equation

$$(13.1) \qquad D^j\,\{2kZ(y) - \mathcal{Q}_0(y) - Z(y)^2 - \varphi(y)\} = O(|y|^{-j-\tilde{\varepsilon}}) \quad (j=0,1,\ \tilde{\varepsilon}>1)\ .$$

Noting that $Z(y)^2 + \varphi(y) = (\text{grad}\,\lambda)^2$, we can rewrite (13.1) as

$$(13.2) \qquad D^j\,\{2k\frac{\partial\lambda}{\partial|y|} - \mathcal{Q}_0(y) - (\text{grad}\,\lambda)^2\} = O(|y|^{-j-\tilde{\varepsilon}}) \quad (j=0,1,\ \tilde{\varepsilon}>1)$$

4° Our proof of Theorems 5.3 and 5.4 is a unification of the ones in Saitō [3] and [6]. The method of the proof has its origin in Jäger [3] in which he treated an differential operator with operator-valued, short-range coefficients.

5° In §5 we defined the stationary modifier by a sort of successive approximation method. The condition $(\tilde{\mathcal{Q}}_0)$ was assumed so that the successive approximation process may be effective. As a result our long-range potential $\mathcal{Q}_0(y)$ is assumed to satisfy the estimates for the derivatives $D^j\mathcal{Q}_0$, $j=0,1,2,\ldots,m$ and m is a rather large number. Here it should be mentioned that our proof of Theorems 5.3 and 5.4 is effective as far as we can construct a stationary modifier which satisfies (1) of Remark 5.9. Hörmander [1] constructed the time-dependent modifier $W(y,t)$ for a type of elliptic operators with more general long-range coefficients than ours. Kitada [2] showed that a stationary modifier satisfying (1) of Remark 5.9 can be constructed by starting with Hörmander's time-dependent modifier. If we use Kitada's stationary modifier we can replace $(\tilde{\mathcal{Q}}_0)$ by

$(\hat{\mathcal{Q}}_0)$. There exist constants C_0 and $0 < \varepsilon \leqq 1$ such that $\mathcal{Q}_0(y)$ is a C^4 function and

$$|D^j\mathcal{Q}_0(y)| \leqq C_0(1+|y|)^{-d(j)} \qquad (y\in\mathbf{R}^N, j=0,1,2,3,4) \ ,$$

where D^j denote an arbitrary derivative of j-th order and $d(j) = j+\varepsilon_0$ $(j=1,2,3)$, $d(4) > 0$, $d(1) + d(4) > 5$.

But in this lecture I adopted the primitive method of successive approximation, because we need further preparations with respect to the theory of partial differential equation in order to introduce the more minute method which starts with Hörmander [1].

6° Theorem 8.7 was first stated and proved explicitly by Kitada [2].

7° Ikebe [4] gave the proof of the orthogonality of the generalized Fourier transforms by treating the Schrödinger operator directly and making use of the Lippmann-Schwinger equation. Our proof of the orthogonality of $F_{\pm}$ is different in using Proposition 8.5 instead of the Lippmann-Schwinger equation.

8° The modified wave operators. The time-dependent modified wave operators $W_{D,\pm}$ for the Schrödinger operator with a long-range potential we defined by Alsholm-Kato [1], Alsholm [1] and Buslaev-Matveev [1] as

(13.3) $$W_{D,\pm} = \underset{t\to\pm\infty}{\text{s-lim}}\, e^{itH}e^{-itH_0 - iX_t},$$

where X_t is a function of H_0. On the other hand from the viewpoint of the stationary method the stationary wave operator $\tilde{W}_{D,\pm}$ should be defined by

(13.4) $$\tilde{W}_{D,\pm} = \tilde{F}_{\pm}^{*}\,\tilde{F}_{0,\pm} \;,$$

$\tilde{F}_{0,\pm}$ being the generalized Fourier transforms associated with H_0. From the orthogonality of the generalized Fourier transforms the completeness of $\tilde{W}_{D,\pm}$ follows immediately. Recently the relation $\tilde{W}_{D,\pm} = \tilde{W}_{D,\pm}$ is shown by Kitada [1], [2], [3] and Ikebe-Isozaki [1], whence follows the completeness of the time-independent modified wave operator $W_{D,\pm}$.

9° In §12 we treated the case that $\mathcal{Q}_1(y)$ is a general short-range potential. Then we approximated $\mathcal{Q}_1(y)$ by a sequence $\{\mathcal{Q}_{1n}(y)\}$, where $\mathcal{Q}_{1n}(y)$ has compact support in $\mathbf{R}^N$. But there is another method which starts with the relations

(13.5) $$(L-k^2)^{-1} = (L_1-k^2)^{-1}\,\{1-C_1(L-k^2)^{-1}\}$$

where $L_1 = -\dfrac{d^2}{dr^2} + B(r) + C_0(r)$, 1 is the identity operator and

$v = (L-k^2)^{-1}f$ means the radiative function for $\{L,k,\ell[f]\}$. Then $F(k)$ can be defined by

$$(13.6) \qquad F(k) = F_1(k)\ \{1-C_1(L-k^2)^{-1}\}$$

This method was adopted in Ikebe [3].

10° In Theorem 11.9 we introduced the generalized Fourier transforms $\tilde{F}_{\pm}$ associated with H. Let $F_{0,\pm}$ be the generalized Fourier transforms associated with H_0, the self-adjoint realization of $-\Delta$. In this case the Green kernel $G_0(r,s,k)$ can be represented by the use of the Hankel function and the exact form of $F_{0,\pm}$ is known as

$$(13.7) \qquad (F_{0,\pm}\Phi)(\xi) = C_I(N)(2\pi)^{-\frac{N}{2}}\ \underset{R\to\infty}{\mathrm{l.i.m.}} \int_{|y|<R} e^{\mp iy\xi}\Phi(y)dy$$

in $L_2(\mathbf{R}^N, d\xi)$ where

$$(13.8) \qquad C_I(N) = -e^{\mp\frac{\pi}{4}(N-1)i}$$

For proof see Saitō [2], §7. This means that $\tilde{F}_{0,\pm}$ are essentially the usual Fourier transforms and the $\tilde{F}_{\pm}$ are the generalization of the usual Fourier transforms in this sense.

11° As was stated in the Introduction, an oscillating long-range potential $Q_0(y)$ such as $Q_0(y) = \frac{1}{|y|}\sin|y|$ does not satisfy any assumption. As for the Schrödinger operator with an oscillating long-range potential we can refer to Mochizuki-Uchiyama [1], [2].

12° Finally let us give two remarks on the condition (5.2) in Assumption 5.1 on a long-range potential Q_0. In order to give a unified treatment for $0 < \varepsilon \leqq 1$ we assumed the condition (5.2). But in the case of $\frac{1}{2} < \varepsilon \leqq 1$ we can adopt a weaker condition $m_0 = 2$. In fact in this case we have

$$(13.9) \qquad Z(y) = \frac{1}{2k} \int_0^r \mathcal{Q}_0(t\omega)dt \qquad (y = r\omega),$$

and the first condition of (5.41) can be weakened as

$$(13.10) \qquad |(D^j Z)(y)| \leqq C(1+|y|)^{-j-\varepsilon} \qquad (j=0,1,2),$$

because the condition

$$(13.11) \qquad |D^3 Z(y)| \leqq C(1+|y|)^{-3-\varepsilon}$$

is used to estimate the term $((Z' + P)u, Bu)_X$ in the proof of Lemma 6.5 only and we can directly estimate this term without making use of (13.11) in the case of $\frac{1}{2} < \varepsilon \leqq 1$. Thus (5.2) can be replaced by

$$(13.12) \qquad m_0 > \frac{2}{\varepsilon} - 1 \quad (0 < \varepsilon \leqq \tfrac{1}{2}) \quad \text{and} \quad m_0 = 2 \quad (\tfrac{1}{2} < \varepsilon \leqq 1).$$

Next let us consider the case the $\mathcal{Q}_0(y)$ is spherically symmetric, i.e. $\mathcal{Q}_0(y) = \mathcal{Q}_0(|y|)$. In this case the stationary modifier $Z(y)$ is also spherically symmetric and the operator M, the functions $\varphi(y;\lambda)$, $P(y;\lambda)$ are all identically zero. Therefore the proof of Theorem 5.4 becomes much simpler. For example, we do not need Lemma 6.6. Moreover (5.35), by which $Z(y)$ is defined, takes the following simpler form:

$$(13.13) \qquad \begin{cases} Z_1(y) = \frac{1}{2k} \mathcal{Q}_0(y), \\ Z_{n+1}(y) = \frac{1}{2k}\{\mathcal{Q}_0(y) + (Z_n(y))^2\}, \end{cases}$$

and we set $Z(y) = Z_{n_0}$ where n_0 is the least integer such that

$(n_0 + 1)\varepsilon > 1$. Noting that the right-hand side of (13.13) does not contain any derivative of $Z_n(y)$, we can replace (5.2) by

(13.14) $$m_0 = 2 \quad (\text{if } \mathcal{Q}_0(y) \text{ is spherically symmetric}).$$

REFERENCES

S. Agmon

[1] Spectral properties of Schrödinger operators and scattering theory, Ann. Scuola Nor. Sup. Pisa (4) 2 (1975), 151-218.

P.K. Alsholm

[1] Wave operators for long-range scattering, Thesis, UC Berkeley (1972).

P.K. Alsholm and T. Kato

[1] Scattering with long-range potentials, Proc. Symp. Pure Math. Vol XIII (1973), 393-399.

N. Aronszajn

[1] A unique continuation theorem for solutions of elliptic partial differential equations or inequalities of second order, J. de Math. 36 (1957), 235-247.

E. Buslaev and V. B. Matveev

[1] Wave operators for the Schrödinger equations, with a slowly decreasing potential, Theoret. and Math. Phys. 2(1970), 266-274.

J.D. Dollard

[1] Asymptotic convergence and the Coulomb interaction, J. Mathematical Phys. 5 (1964), 729-738.

A. Erdélyi and others

[1] Higher transcendental function II, McGraw-Hill, New York, 1953.

L. Hörmander

[1] The existence of wave operators in scattering theory; Math. Z. 149(1976), 69-91.

T. Ikebe

[1] Eigenfunction expansions associated with the Schrödinger operators and their application to scattering theory, Arch. Rational Mech. Anal. 5(1960), 1-34.

[2] Spectral representations for the Schrödinger operators with long-range potentials, J. Functional Analysis 20 (1975), 158-177.

[3] Spectral representations for the Schrödinger operators with long-range potentials, II, Publ. Res. Inst. Math. Sci. Kyoto Univ. 11 (1976), 551-558.

[4] Remarks on the orthogonality of eigenfunctions for the Schrödinger operator in $\mathbb{R}^n$ J. Fac. Sci. Univ. Tokyo 17 (1970), 355-361.

T. Ikebe and H. Isozaki

[1] Completeness of modified wave operators for long-range potentials, Preprint.

T. Ikebe and T. Kato

[1] Uniqueness of the self-adjoint extension of singular elliptic differential operators, Arch. Ratinal Mech. Anal. 9 (1962), 77-92.

T. Ikebe and Y. Saitō

[1] Limiting absorption method and absolute continuity for the Schrödinger operator, J. Math. Kyoto Univ. 7(1972), 513-542.

W. Jäger

[1] Über das Dirichletsche Aussenraumproblem für die Schwingungsgleichung, Math. Z. 95 (1967), 299-323.

[2] Zur Theorie der Schwingungsgleichung mit variablen Koeffizienten in Aussengebieten, Math. Z. 102 (1967). 62-88.

[3] Das asymptotische Verhalten von Lösungen eines typus von Differential gleichungen, Math. Z. 112 (1969), 26-36.

[4] Ein gewöhnlicher Differentialoperator zweiter Ordnung für Funktionen mit Werten in einem Hilbertraum, Math. Z. 113 (1970) 68-98.

T. Kato

[1] Perturbation Theory for Linear Operators, Second Edition, Springer, Berlin, 1976.

H. Kitada

[1] On the completeness of modified wave operators, Proc. Japan. Acad. 52 (1976), 409-412.

[2] Scattering theory for Schrödinger operators with long-range potentials, I, J. Math. Soc. Japan 29 (1977), 665-691.

[3] Scattering theory for Schrödinger operators with long-range potentials, II, J. Math. Soc. Japan 30 (1978), 603-632.

S. T. Kuroda

[1] Construction of eigenfunction expansions by the perturbation method and its application to n-dimensional Schrödinger operators, MRC Technical Summary Report No. 744 (1967).

R. Lavine

[1] Absolute continuity of positive spectrum for Schrödinger operators with long-range potentials, J. Functional Analysis 12 (1973), 30-54.

K. Mochizuki and J. Uchiyama

[1] Radiation conditions and spectral theory for 2-body Schrödinger operators with "oscillatory" long-range potentials, I, the principle of limiting absorption, to appear in J. Math. Kyoto Univ.

[2] Radiation conditions and spectral theory for 2-body Schrödinger operators with "oscillatory" long-range potentials, II, spectral representation, to appear in J. Math. Kyoto Univ.

Y. Saitō

[1] The principle of limiting absorption for second-order differential equations with operator-valued coefficients, Publ. Res. Inst. Math. Sci. Kyoto Univ. 7 (1971/72), 581-619.

[2] Spectral and scattering theory for second-order differential operators with operator-valued coefficients, Osaka J. Math. 9(1972), 463-498.

[3] Spectral theory for second-order differential operators with long-range operator-valued coefficients, I, Japan J. Math. New Ser. 1 (1975), 311-349.

[4] Spectral theory for second-order differential operators with long-range operator-valued coefficients, II, Japan J. Math. New Ser. 1 (1975), 351-382.

[5] Eigenfunction expansions for differential operators with operator-valued coefficients and their applications to the Schrödinger operators with long-range potentials, International Symposium on Mathematical Problems in Theoretical Physics. Proceedings 1975, Lecture Notes in Physics 39(1975), 476-482, Springer.

[6] On the asymptotic behavior of the solutions of the Schrödinger equation $(-\Delta + Q(y)-k^2)V = F$, Osaka J. Math. 14(1977), 11-35.

[7] Eigenfunction expansions for the Schrödinger operators with long-range potentials $Q(y) = O(|y|^{-\varepsilon})$, $\varepsilon > 0$, Osaka J. Math. (1977), 37-53.

D. W. Thoe

[1] Eigenfunction expansions associated with Schrödinger operators in $\mathbb{R}^n$, $n \geqq 4$, Arch. Rational Mech. Anal. 26(1967), 335-356.

K. Yosida

[1] Functional Analysis, Springer, 1964.

NOTATION INDEX

SUBJECT INDEX

Vol. 551: Algebraic K-Theory, Evanston 1976. Proceedings. Edited by M. R. Stein. XI, 409 pages. 1976.

Vol. 552: C. G. Gibson, K. Wirthmüller, A. A. du Plessis and E. J. N. Looijenga. Topological Stability of Smooth Mappings. V, 155 pages. 1976.

Vol. 553: M. Petrich, Categories of Algebraic Systems. Vector and Projective Spaces, Semigroups, Rings and Lattices. VIII, 217 pages. 1976.

Vol. 554: J. D. H. Smith, Mal'cev Varieties. VIII, 158 pages. 1976.

Vol. 555: M. Ishida, The Genus Fields of Algebraic Number Fields. VII, 116 pages. 1976.

Vol. 556: Approximation Theory. Bonn 1976. Proceedings. Edited by R. Schaback and K. Scherer. VII, 466 pages. 1976.

Vol. 557: W. Iberkleid and T. Petrie, Smooth S^1 Manifolds. III, 163 pages. 1976.

Vol. 558: B. Weisfeiler, On Construction and Identification of Graphs. XIV, 237 pages. 1976.

Vol. 559: J.-P. Caubet, Le Mouvement Brownien Relativiste. IX, 212 pages. 1976.

Vol. 560: Combinatorial Mathematics, IV, Proceedings 1975. Edited by L. R. A. Casse and W. D. Wallis. VII, 249 pages. 1976.

Vol. 561: Function Theoretic Methods for Partial Differential Equations. Darmstadt 1976. Proceedings. Edited by V. E. Meister, N. Weck and W. L. Wendland. XVIII, 520 pages. 1976.

Vol. 562: R. W. Goodman, Nilpotent Lie Groups: Structure and Applications to Analysis. X, 210 pages. 1976.

Vol. 563: Séminaire de Théorie du Potentiel. Paris, No. 2. Proceedings 1975–1976. Edited by F. Hirsch and G. Mokobodzki. VI, 292 pages. 1976.

Vol. 564: Ordinary and Partial Differential Equations, Dundee 1976. Proceedings. Edited by W. N. Everitt and B. D. Sleeman. XVIII, 551 pages. 1976.

Vol. 565: Turbulence and Navier Stokes Equations. Proceedings 1975. Edited by R. Temam. IX, 194 pages. 1976.

Vol. 566: Empirical Distributions and Processes. Oberwolfach 1976. Proceedings. Edited by P. Gaenssler and P. Révész. VII, 146 pages. 1976.

Vol. 567: Séminaire Bourbaki vol. 1975/76. Exposés 471–488. IV, 303 pages. 1977.

Vol. 568: R. E. Gaines and J. L. Mawhin, Coincidence Degree, and Nonlinear Differential Equations. V, 262 pages. 1977.

Vol. 569: Cohomologie Etale SGA 4½. Séminaire de Géométrie Algébrique du Bois-Marie. Edité par P. Deligne. V, 312 pages. 1977.

Vol. 570: Differential Geometrical Methods in Mathematical Physics, Bonn 1975. Proceedings. Edited by K. Bleuler and A. Reetz. VIII, 576 pages. 1977.

Vol. 571: Constructive Theory of Functions of Several Variables, Oberwolfach 1976. Proceedings. Edited by W. Schempp and K. Zeller. VI. 290 pages. 1977

Vol. 572: Sparse Matrix Techniques, Copenhagen 1976. Edited by V. A. Barker. V, 184 pages. 1977.

Vol. 573: Group Theory, Canberra 1975. Proceedings. Edited by R. A. Bryce, J. Cossey and M. F. Newman. VII, 146 pages. 1977.

Vol. 574: J. Moldestad, Computations in Higher Types. IV, 203 pages. 1977.

Vol. 575: K-Theory and Operator Algebras, Athens, Georgia 1975. Edited by B. B. Morrel and I. M. Singer. VI, 191 pages. 1977.

Vol. 576: V. S. Varadarajan, Harmonic Analysis on Real Reductive Groups. VI, 521 pages. 1977.

Vol. 577: J. P. May, E_∞ Ring Spaces and E_∞ Ring Spectra. IV, 268 pages. 1977.

Vol. 578: Séminaire Pierre Lelong (Analyse) Année 1975/76. Edité par P. Lelong. VI, 327 pages. 1977.

Vol. 579: Combinatoire et Représentation du Groupe Symétrique, Strasbourg 1976. Proceedings 1976. Edité par D. Foata. IV, 339 pages. 1977.

Vol. 580: C. Castaing and M. Valadier, Convex Analysis and Measurable Multifunctions. VIII, 278 pages. 1977.

Vol. 581: Séminaire de Probabilités XI, Université de Strasbourg. Proceedings 1975/1976. Edité par C. Dellacherie, P. A. Meyer et M. Weil. VI, 574 pages. 1977.

Vol. 582: J. M. G. Fell, Induced Representations and Banach *-Algebraic Bundles. IV, 349 pages. 1977.

Vol. 583: W. Hirsch, C. C. Pugh and M. Shub, Invariant Manifolds. IV, 149 pages. 1977.

Vol. 584: C. Brezinski, Accélération de la Convergence en Analyse Numérique. IV, 313 pages. 1977.

Vol. 585: T. A. Springer, Invariant Theory. VI, 112 pages. 1977.

Vol. 586: Séminaire d'Algèbre Paul Dubreil, Paris 1975–1976 (29ème Année). Edited by M. P. Malliavin. VI, 188 pages. 1977.

Vol. 587: Non-Commutative Harmonic Analysis. Proceedings 1976. Edited by J. Carmona and M. Vergne. IV, 240 pages. 1977.

Vol. 588: P. Molino, Théorie des G-Structures: Le Problème d'Equivalence. VI, 163 pages. 1977.

Vol. 589: Cohomologie l-adique et Fonctions L. Séminaire de Géométrie Algébrique du Bois-Marie 1965–66, SGA 5. Edité par L. Illusie. XII, 484 pages. 1977.

Vol. 590: H. Matsumoto, Analyse Harmonique dans les Systèmes de Tits Bornologiques de Type Affine. IV, 219 pages. 1977.

Vol. 591: G. A. Anderson, Surgery with Coefficients. VIII, 157 pages. 1977.

Vol. 592: D. Voigt, Induzierte Darstellungen in der Theorie der endlichen, algebraischen Gruppen. V, 413 Seiten. 1977.

Vol. 593: K. Barbey and H. König, Abstract Analytic Function Theory and Hardy Algebras. VIII, 260 pages. 1977.

Vol. 594: Singular Perturbations and Boundary Layer Theory, Lyon 1976. Edited by C. M. Brauner, B. Gay, and J. Mathieu. VIII, 539 pages. 1977.

Vol. 595: W. Hazod, Stetige Faltungshalbgruppen von Wahrscheinlichkeitsmaßen und erzeugende Distributionen. XIII, 157 Seiten. 1977.

Vol. 596: K. Deimling, Ordinary Differential Equations in Banach Spaces. VI, 137 pages. 1977.

Vol. 597: Geometry and Topology, Rio de Janeiro, July 1976. Proceedings. Edited by J. Palis and M. do Carmo. VI, 866 pages. 1977.

Vol. 598: J. Hoffmann-Jørgensen, T. M. Liggett et J. Neveu, Ecole d'Eté de Probabilités de Saint-Flour VI – 1976. Edité par P.-L. Hennequin. XII, 447 pages. 1977.

Vol. 599: Complex Analysis, Kentucky 1976. Proceedings. Edited by J. D. Buckholtz and T. J. Suffridge. X, 159 pages. 1977.

Vol. 600: W. Stoll, Value Distribution on Parabolic Spaces. VIII, 216 pages. 1977.

Vol. 601: Modular Functions of one Variable V, Bonn 1976. Proceedings. Edited by J.-P. Serre and D. B. Zagier. VI, 294 pages. 1977.

Vol. 602: J. P. Brezin, Harmonic Analysis on Compact Solvmanifolds. VIII, 179 pages. 1977.

Vol. 603: B. Moishezon, Complex Surfaces and Connected Sums of Complex Projective Planes. IV, 234 pages. 1977.

Vol. 604: Banach Spaces of Analytic Functions, Kent, Ohio 1976. Proceedings. Edited by J. Baker, C. Cleaver and Joseph Diestel. VI, 141 pages. 1977.

Vol. 605: Sario et al., Classification Theory of Riemannian Manifolds. XX, 498 pages. 1977.

Vol. 606: Mathematical Aspects of Finite Element Methods. Proceedings 1975. Edited by I. Galligani and E. Magenes. VI, 362 pages. 1977.

Vol. 607: M. Métivier, Reelle und Vektorwertige Quasimartingale und die Theorie der Stochastischen Integration. X, 310 Seiten. 1977.

Vol. 608: Bigard et al., Groupes et Anneaux Réticulés. XIV, 334 pages. 1977.